*Boundary value problems governed
by second order elliptic systems*

TO GAELLE

Boundary Value Problems Governed by Second Order Elliptic Systems

David L. Clements
University of Adelaide

Pitman Advanced Publishing Program

Boston · London · Melbourne

PITMAN PUBLISHING LIMITED
39 Parker Street, London WC2B 5PB

PITMAN PUBLISHING INC.
1020 Plain Street, Marshfield, Massachusetts

Associated Companies
Pitman Publishing Pty Ltd, Melbourne
Pitman Publishing New Zealand Ltd, Wellington
Copp Clark Pitman, Toronto

First published 1981

AMS Subject Classifications: (main) 35J55, 73-02, 65N99
                                          (subsidiary) 73C30, 73J05, 73M05

Library of Congress Cataloging in Publication Data

Clements, David L
  Boundary value problems governed by second order elliptic systems.

  (Monographs and studies in mathematics; 12)
  Bibliography: p.
  Includes indexes.
  1. Boundary value problems. 2. Differential equations, Elliptic. I. Title. II. Series.
QA379.C53   515.3′5   80-20820
ISBN 0-273-08502-6

Typeset in Northern Ireland at The Universities Press (Belfast) Ltd.
Printed and bound in Great Britain at The Pitman Press, Bath

Contents

Preface

This book is concerned with the solution of boundary value problems governed by a system of second order elliptic partial differential equations. It is not intended to be a comprehensive account of the subject but rather a description of the derivation of some of the solutions which have been found to be useful in applications over recent years. Essentially, the book provides a connected account of the derivation and application of these solutions. In addition to collecting together some of the published work, this volume contains a substantial amount of material which has not previously been published. In particular, this new material includes some of the results in Chapters 1 and 3 and many of the results involving the derivation and application of boundary integral equations for the numerical solution of particular problems.

The second order system of equations is introduced in Chapter 1 and a general solution obtained in terms of arbitrary analytic functions of a complex variable. Various properties of the constants appearing in the general solutions are examined and, in particular, the simplifications which occur for particular classes of elliptic systems are noted in some detail.

Chapter 2 contains an account of some of the important physical problem areas which are governed by the elliptic system. It is noted, in passing, that Laplace's equation is a special case of the system and hence all the solutions obtained in this book are applicable to relevant problems governed by Laplace's equation. Also, it is noted that the equations governing generalized plane thermostatics and elastostatics for anisotropic materials are obtainable as special cases of the elliptic system under consideration. This is a particularly important area of application since it is now widely recognized that some insight into the behaviour of certain classes of fibre-reinforced materials may be obtained by modelling these materials as homogeneous anisotropic materials. Hence, in recent years, considerable interest has centred on the solution of boundary-value problems for anisotropic materials.

In Chapter 3 both complex variable and integral transform techniques are employed to obtain the solution to a number of particular boundary value problems. Chapter 4 contains applications of these solutions to a number of problems in elasticity and thermostatics. In many of these solutions attention is drawn to the effect of strong anisotropy since this is of importance in the study of fibre-reinforced materials. Specifically, it is possible to use these solutions to identify differences in behaviour between strongly anisotropic materials and those exhibiting normal anisotropy. This is of considerable interest since it is the assumption that strongly anisotropic materials will largely behave in a similar manner to those exhibiting ordinary anisotropy that has led to a number of difficulties with fibre-reinforced materials.

Chapters 5 and 6 deal with the derivation of various boundary integral equations. These equations permit the solution to a general boundary value problem for the elliptic system to be expressed in terms of an integral taken round the boundary of the region under consideration. In general, this integral cannot be evaluated analytically and hence the boundary integral formulae essentially form the basis of an effective numerical procedure for the solution of a particular problem. Various different boundary integral equations are presented. The first may be used to obtain the numerical solution to a well-posed problem for a region with an arbitrary boundary. The remainder are for regions with particular types of boundaries and hence have a more restricted application. However, when they are applicable they will generally be more useful than the integral equation for the general problem.

The final chapter is devoted to the application of the integral equations derived in Chapter 6 to the solution of particular boundary-value problems. Various problems are considered with the main aim being to illustrate how each of the different boundary integral equations may be employed to solve effectively the relevant class of problems.

1

The second order system of equations

1.1 Statement of the boundary-value problem

Consider the system of elliptic partial differential equations

$$a_{ijkl}\frac{\partial^2 \phi_k}{\partial x_j \partial x_l} = 0 \quad \text{for} \quad i = 1, 2, \ldots, N \tag{1.1.1}$$

in which ϕ_k for $k = 1, 2, \ldots, N$ are complex-valued functions of the dependent variables x_1 and x_2 and where the summation convention is implied for all lower case italic subscripts. Although the ϕ_k are taken to be complex-valued functions they will be real-valued in most of the cases considered. The a_{ijkl} occurring in (1.1.1) are assumed to be real constants which satisfy the symmetry condition

$$a_{ijkl} = a_{klij}. \tag{1.1.2}$$

Further, since the system (1.1.1) is elliptic the constants a_{ijkl} must be such that

$$a_{ijkl}\xi_{ij}\xi_{kl} > 0 \tag{1.1.3}$$

for every non-zero $N \times 2$ real matrix $[\xi_{ij}]$. It is required to find a solution to (1.1.1) valid in a region R in E^2 with boundary C which consists of a finite number of piecewise smooth closed curves. On C either the dependent variables ϕ_k are specified or the P_i are specified where

$$P_i = a_{ijkl}\frac{\partial \phi_k}{\partial x_l} n_j \quad \text{for} \quad i = 1, 2, \ldots, N, \tag{1.1.4}$$

where n_j is the unit (outward) normal to R. Now from Green's theorem it follows that

$$\int_R a_{ijkl}\frac{\partial^2 \phi_k}{\partial x_j \partial x_l} \, dR = \int_C a_{ijkl}\frac{\partial \phi_k}{\partial x_l} n_j \, ds$$

1

so that, using (1.1.4)

$$\int_R a_{ijkl} \frac{\partial^2 \phi_k}{\partial x_j\, \partial x_l}\, dR = \int_C P_i\, ds. \tag{1.1.5}$$

From (1.1.1) it is apparent that the left hand side of (1.1.5) is zero. Hence if P_i is specified over all of C then it must be such that

$$\int_C P_i\, ds = 0. \tag{1.1.6}$$

1.2   Solution of the system in terms of analytic functions

Consider the possibility of finding a solution to (1.1.1) with the form

$$\phi_k(x_1, x_2) = A_k f(x_1 + \tau x_2), \tag{1.2.1}$$

where f is an analytic function of a complex variable and τ, $A_1, \ldots, A_n$ are complex constants which will be determined during the solution process. Substitution of (1.2.1) into (1.1.1) yields the linear algebraic system of N homogeneous equations in N unknowns

$$(a_{i1k1} + a_{i1k2}\tau + a_{i2k1}\tau + a_{i2k2}\tau^2)A_k = 0. \tag{1.2.2}$$

For this system to have a non-trivial solution it is necessary that the determinant of coefficients be zero. Hence

$$|a_{i1k1} + a_{i1k2}\tau + a_{i2k1}\tau + a_{i2k2}\tau^2| = 0. \tag{1.2.3}$$

This determinant is clearly a polynomial of degree $2N$ in τ. It has $2N$ roots which will be required to be distinct. (From a numerical viewpoint equal roots may be readily taken as a limiting case of distinct roots). As will be shown in Section 1.3 the $2N$ roots are of necessity complex since the ellipticity condition (1.1.3) will be violated for a real root. Since the coefficients a_{ijkl} are real it thus follows that the roots will occur in complex conjugate pairs. The roots with positive imaginary part will be denoted by τ_α, $\alpha = 1, 2, \ldots, N$ with conjugates denoted by $\bar{\tau}_\alpha$. The corresponding A_k will be denoted by $A_{k\alpha}$ and $\bar{A}_{k\alpha}$, respectively. Real solutions of (1.1.1) may therefore be written in the form

$$\phi_k = \sum_\alpha A_{k\alpha} f_\alpha(z_\alpha) + \sum_\alpha \bar{A}_{k\alpha} \bar{f}_\alpha(\bar{z}_\alpha), \tag{1.2.4}$$

where $z_\alpha = x_1 + \tau_\alpha x_2$, the sums are taken from 1 to N, $f_1, f_2, \ldots, f_N$ are arbitrary analytic functions and

$$\bar{f}_\alpha(z) = \overline{f_\alpha(\bar{z})} \quad \text{for} \quad \alpha = 1, 2, \ldots, N.$$

In order to obtain a representation for P_i in terms of the arbitrary analytic functions $f_\alpha(z)$ it is convenient to introduce the new variables ψ_{ij} defined by

$$\psi_{ij} = a_{ijkl} \frac{\partial \phi_k}{\partial x_l} \quad \text{for} \quad i = 1, 2, \ldots, N. \tag{1.2.5}$$

so that, from (1.1.4),

$$P_i = \psi_{ij} n_j. \tag{1.2.6}$$

Substitution of (1.2.4) into (1.2.5) yields

$$\psi_{ij} = \sum_\alpha L_{ij\alpha} f'_\alpha(z_\alpha) + \sum_\alpha \bar{L}_{ij\alpha} \bar{f}'_\alpha(\bar{z}_\alpha), \tag{1.2.7}$$

where

$$L_{ij\alpha} = (a_{ijk1} + \tau_\alpha a_{ijk2}) A_{k\alpha} \tag{1.2.8}$$

and primes denote differentiation with respect to the argument in question. From (1.1.2), (1.2.2) and (1.2.8) it follows that

$$L_{i1\alpha} = -\tau_\alpha L_{i2\alpha}. \tag{1.2.9}$$

Hence, in general it will only be necessary to consider the $N \times N$ matrix $[L_{i2\alpha}]$. The matrix $[L_{i1\alpha}]$ may then be determined through (1.2.9).

It will be useful subsequently to have some alternative forms for equations (1.2.4) and (1.2.5). Let

$$\sum_\alpha A_{k\alpha} f_\alpha(z) = \theta_k(z), \tag{1.2.10}$$

where the $\theta_k(z)$, $k = 1, 2, \ldots, N$ are analytic functions of the complex variable z. It will be shown in Section 1.3 that the $N \times N$ matrix $[A_{k\alpha}]$ is non-singular and hence, from (1.2.10),

$$f_\alpha(z) = N_{\alpha j} \theta_j(z), \tag{1.2.11}$$

with the matrix $[N_{\alpha j}]$ defined by

$$\sum_\alpha A_{k\alpha} N_{\alpha j} = \delta_{ij}, \tag{1.2.12}$$

where δ_{ij} is the Kronecker delta. Substitution of (1.2.10) into (1.2.4) and (1.2.7) yields

$$\phi_k = \sum_\alpha A_{k\alpha} N_{\alpha j} \theta_j(z_\alpha) + \sum_\alpha \bar{A}_{k\alpha} \bar{N}_{\alpha j} \bar{\theta}_j(\bar{z}_\alpha), \tag{1.2.13}$$

$$\psi_{ij} = \sum_\alpha L_{ij\alpha} N_{\alpha k} \theta'_k(z_\alpha) + \sum_\alpha \bar{L}_{ij\alpha} \bar{N}_{\alpha k} \bar{\theta}'_k(\bar{z}_\alpha). \tag{1.2.14}$$

In particular, on $x_2 = 0$ (1.2.13) and (1.2.14) become (by virtue of (1.2.12))

$$\phi_k = \theta_k(x_1) + \bar{\theta}_k(x_1), \tag{1.2.15}$$

$$\psi_{i2} = C_{ik}\theta'_k(x_1) + \bar{C}_{ik}\bar{\theta}'_k(x_1), \tag{1.2.16}$$

where

$$C_{ik} = \sum_\alpha L_{i2\alpha}N_{\alpha k}. \tag{1.2.17}$$

A second useful alternative representation for ϕ_k and ψ_{ij} is obtained by putting

$$\sum_\alpha L_{i2\alpha}f_\alpha(z) = \chi_i(z), \tag{1.2.18}$$

where the $\chi_i(z)$, $i = 1, 2, \ldots, N$ are analytic functions of the complex variable z. The $N \times N$ matrix $[L_{i2\alpha}]$ is non-singular (see Section 1.3) so that

$$f_\alpha(z) = M_{\alpha j}\chi_j(z), \tag{1.2.19}$$

where

$$\sum_\alpha L_{i2\alpha}M_{\alpha j} = \delta_{ij}. \tag{1.2.20}$$

Substitution of (1.2.19) into (1.2.4) and (1.2.7) now yields

$$\phi_k = \sum_\alpha A_{k\alpha}M_{\alpha j}\chi_j(z_\alpha) + \sum_\alpha \bar{A}_{k\alpha}\bar{M}_{\alpha j}\bar{\chi}_j(\bar{z}_\alpha), \tag{1.2.21}$$

$$\psi_{ij} = \sum_\alpha L_{ij\alpha}M_{\alpha k}\chi'_k(z_\alpha) + \sum_\alpha \bar{L}_{ij\alpha}\bar{M}_{\alpha k}\bar{\chi}'_k(\bar{z}_\alpha). \tag{1.2.22}$$

On the plane $x_2 = 0$ (1.2.21) and (1.2.22) reduce to

$$\phi_k = B_{kj}\chi_j(x_1) + \bar{B}_{kj}\bar{\chi}_j(x_1), \tag{1.2.23}$$

$$\psi_{i2} = \chi'_k(x_1) + \bar{\chi}'_k(x_1), \tag{1.2.24}$$

where

$$B_{kj} = \sum_\alpha A_{k\alpha}M_{\alpha j}. \tag{1.2.25}$$

1.3 Properties of the constants occurring in the basic equations

In this section some properties of the constants occurring in the basic equations are derived.

Theorem 1.3.1 *The equation*

$$|a_{i1k1} + a_{i1k2}\tau + a_{i2k1}\tau + a_{i2k2}\tau^2| = 0 \tag{1.3.1}$$

has only complex roots.

Proof The proof is a minor extension of a result established by Eshelby *et al.* [18].

Suppose that τ' is a real root of (1.3.1). Then the corresponding non-trivial A_k, found from (1.2.2) will be real. Now multiplication of (1.2.2) by A_i yields

$$(a_{i1k1} + a_{i1k2}\tau' + a_{i2k1}\tau' + a_{i2k2}\tau'^2)A_k A_i = 0.$$

Thus the choice

$$\frac{\partial \phi_k}{\partial x_1} = A_k, \qquad \frac{\partial \phi_k}{\partial x_2} = A_k \tau'$$

yields

$$a_{ijkl}\frac{\partial \phi_i}{\partial x_j}\frac{\partial \phi_k}{\partial x_l} = 0$$

which violates the ellipticity condition (1.1.3). Hence the assumption that τ' is real is unacceptable.

The next two theorems are modifications and extensions of results established by Stroh [38].

Theorem 1.3.2 *The matrix $\bar{A}_{i\beta}L_{i2\alpha}$ has skew-Hermitian symmetry.*

Proof From (1.2.8) it follows that

$$\bar{\tau}_\beta \bar{A}_{i\beta}L_{i2\alpha} = \bar{\tau}_\beta \bar{A}_{i\beta}a_{i2k1}A_{k\alpha} + \bar{\tau}_\beta \bar{A}_{i\beta}a_{i2k2}\tau_\alpha A_{k\alpha}. \tag{1.3.2}$$

Now use of (1.2.2) in (1.2.8) yields

$$L_{i2\alpha} = -(\tau_\alpha^{-1}a_{i1k1} + a_{i1k2})A_{k\alpha} \tag{1.3.3}$$

and hence

$$\bar{A}_{i\beta}L_{i2\alpha}\tau_\alpha = -\bar{A}_{i\beta}a_{i1k1}A_{k\alpha} - \bar{A}_{i\beta}a_{i1k2}\tau_\alpha A_{k\alpha}. \tag{1.3.4}$$

Subtraction of (1.3.4) from (1.3.2) gives

$$\bar{A}_{i\beta}L_{i2\alpha}(\bar{\tau}_\beta - \tau_\alpha) = \bar{A}_{i\beta}a_{i1k1}A_{k\alpha} + \bar{\tau}_\beta \bar{A}_{i\beta}a_{i2k2}\tau_\alpha A_{k\alpha}$$
$$+ (\bar{\tau}_\beta \bar{A}_{i\beta}a_{i2k1}A_{k\alpha} + \bar{A}_{i\beta}a_{i1k2}\tau_\alpha A_{k\alpha}). \tag{1.3.5}$$

Each of the terms on the right hand side of (1.3.5) has Hermitian symmetry, and hence the left hand side must also; that is

$$\bar{A}_{i\beta}L_{i2\alpha}(\bar{\tau}_\beta - \tau_\alpha) = A_{i\alpha}\bar{L}_{i2\beta}(\tau_\alpha - \bar{\tau}_\beta) \tag{1.3.6}$$

and since $(\bar{\tau}_\beta - \tau_\alpha)$ has negative imaginary part and hence is not zero it immediately follows that

$$\bar{A}_{i\beta}L_{i2\alpha} = -A_{i\alpha}\bar{L}_{i2\beta} \qquad (1.3.7)$$

which is the required result.

Before proceeding to the next theorem it is convenient to consider further the ellipticity condition (1.1.3). Let $\boldsymbol{\zeta}$, $\boldsymbol{\eta}$ be the $1 \times N$ matrices defined by

$$\boldsymbol{\zeta} = [\xi_{i1}], \qquad \boldsymbol{\eta} = [\xi_{i2}]$$

and $\mathbf{C}_1, \mathbf{C}_2$ and $\mathbf{C}_3$ be the $N \times N$ matrices defined by

$$\mathbf{C}_1 = [a_{i1k1}], \qquad \mathbf{C}_2 = [a_{i1k2}], \qquad \mathbf{C}_3 = [a_{i2k2}].$$

The condition (1.1.3) can now be written in terms of partitioned matrices in the form

$$[\boldsymbol{\zeta} \mid \boldsymbol{\eta}]\begin{bmatrix} \mathbf{C}_1 & \mathbf{C}_2 \\ \hline \mathbf{C}_2^T & \mathbf{C}_3 \end{bmatrix}\begin{bmatrix} \boldsymbol{\zeta}^T \\ \hline \boldsymbol{\eta}^T \end{bmatrix} > 0 \quad \text{for all } [\boldsymbol{\zeta}, \boldsymbol{\eta}] \neq \mathbf{0} \qquad (1.3.8)$$

where T denotes the transpose. The homogeneous quadratic form on the left hand side of (1.3.8) is, by virtue of (1.3.8), positive definite and hence, from the properties of positive definite quadratic forms, it follows that the determinant

$$\begin{vmatrix} \mathbf{C}_1 & \mathbf{C}_2 \\ \hline \mathbf{C}_2^T & \mathbf{C}_3 \end{vmatrix} \qquad (1.3.9)$$

and all its principal minors of all orders are positive.

Theorem 1.3.3   *The matrix $[A_{k\alpha}]$ is non-singular.*

Proof   Suppose the matrix $[A_{k\alpha}]$ is singular. Then the columns of the matrix form a linearly dependent set of vectors and hence there exist scalars ξ_α, $\alpha = 1, 2, \ldots, N$ such that

$$\sum_\alpha \xi_\alpha A_{k\alpha} = 0 \qquad (1.3.10)$$

with not all the ξ_α zero. If (1.3.5) is multiplied by $\xi_\alpha \bar{\xi}_\beta$ and summed over α and β then, using (1.3.7) and (1.3.10) it follows that the left hand side becomes

$$\sum_\alpha \sum_\beta \bar{\xi}_\beta \bar{A}_{i\beta} L_{i2\beta}(\bar{\tau}_\beta - \tau_\alpha)\xi_\alpha = -\sum_\beta \bar{L}_{i2\beta}\bar{\tau}_\beta\bar{\xi}_\beta \sum_\alpha A_{i\alpha}\xi_\alpha$$

$$= -\sum_\beta \bar{A}_{i\beta}\bar{\xi}_\beta \sum_\alpha L_{i2\alpha}\tau_\alpha\xi_\alpha = 0. \qquad (1.3.11)$$

Again using (1.3.10) the right hand side becomes

$$\left(\sum_\beta \bar\xi_\beta \bar\tau_\beta \bar A_{i\beta}\right) a_{i2k2} \left(\sum_\alpha \xi_\alpha \tau_\alpha A_{k\alpha}\right) = 0. \tag{1.3.12}$$

The determinant $|a_{i2k2}|$ is a principal minor of the $2N \times 2N$ determinant (1.3.9) and hence the matrix $[a_{i2k2}]$ must be positive definite. Therefore

$$\sum_\alpha \xi_\alpha \tau_\alpha A_{k\alpha} = 0. \tag{1.3.13}$$

Now from equation (1.2.2)

$$a_{i1k1} \sum_\alpha \xi_\alpha A_{k\alpha} + (a_{i1k2} + a_{21k1}) \sum_\alpha \xi_\alpha \tau_\alpha A_{k\alpha} + a_{i2k2} \sum_\alpha \xi_\alpha \tau_\alpha^2 A_{k\alpha} = 0$$

and, on employing (1.3.10) and (1.3.13) this reduces to

$$a_{i2k2} \sum_\alpha \xi_\alpha \tau_\alpha^2 A_{k\alpha} = 0. \tag{1.3.14}$$

Hence, since $[a_{i2k2}]$ is positive definite

$$\sum_\alpha \xi_\alpha \tau_\alpha^2 A_{k\alpha} = 0. \tag{1.3.15}$$

Repeating the process we obtain the general result

$$\sum_\alpha \xi_\alpha \tau_\alpha^n A_{k\alpha} = 0 \quad \text{for} \quad n = 0, 1, 2, \ldots, N \tag{1.3.16}$$

For a fixed k, (1.3.16) constitutes N simultaneous homogeneous equations for the three unknowns $\xi_\alpha A_{k\alpha}$. The $N \times N$ determinant of these equations is

$$\begin{vmatrix} 1 & 1 & 1 & \ldots & 1 \\ \tau_1 & \tau_2 & \tau_3 & \ldots & \tau_N \\ \tau_1^2 & \tau_2^2 & \tau_3^2 & \ldots & \tau_N^2 \\ & & \cdot & & \\ & & \cdot & & \\ & & \cdot & & \\ \tau_1^N & \tau_2^N & \tau_3^N & \ldots & \tau_N^N \end{vmatrix} = \prod_{\alpha=1}^{N-1} \prod_{\beta=\alpha+1}^{N} (\tau_\beta - \tau_\alpha). \tag{1.3.17}$$

This determinant is not zero since all of the τ_α are required to be distinct. Hence the only solution to the system (1.3.16) is the trivial one $\xi_\alpha A_{k\alpha} = 0$. But $A_{k\alpha}$ is not identically zero so $\xi_\alpha = 0$. Thus (1.3.10) implies that the $\xi_\alpha = 0$ so that $[A_{k\alpha}]$ is non-singular.

Theorem 1.3.4 *The matrix $[L_{i2\alpha}]$ is non-singular.*

Proof   As in the previous theorem, suppose the $L_{i2\alpha}$ are not linearly independent so that there exist scalars ξ_α, not all zero, such that

$$\sum_\alpha \xi_\alpha L_{i2\alpha} = 0. \tag{1.3.18}$$

To establish that this assumption leads to a contradiction, a solution to a particular boundary value problem is constructed. Suppose the region R consists of the half-plane $x_2 < 0$ and on the boundary $x_2 = 0$ of the half-plane P_i is prescribed according to

$$P_i = \begin{cases} P(x_1) & \text{for} \quad i = 1, \\ 0 & \text{for} \quad i = 2, 3, \ldots, N, \end{cases} \tag{1.3.19}$$

where

$$P(x_1) = \begin{cases} -P_0 & \text{for} \quad -2a < x_1 < -a, \\ P_0 & \text{for} \quad -a < x_1 < a, \\ -P_0 & \text{for} \quad a < x_1 < 2a, \end{cases} \tag{1.3.20}$$

where P_0 and a are positive constants. Also $\phi_i \to 0$, $i = 1, 2, \ldots, N$ as $|z| = (x_1^2 + x_2^2)^{1/2} \to \infty$. A solution to this problem may be obtained by taking

$$f_\alpha(z_\alpha) = \eta_\alpha g(z_\alpha) \tag{1.3.21}$$

(where the η_α, $\alpha = 1, 2, \ldots, N$ are constants and $g(z)$ is an analytic function) in (1.2.4) and (1.2.7). From (1.3.20) and (1.2.6) the boundary condition on $x_2 = 0$ may be written in terms of ψ_{i2} as

$$\psi_{12} = \begin{cases} -P_0 & \text{for} \quad -2a < x_1 < -a, \\ P_0 & \text{for} \quad -a < x_1 < a, \\ -P_0 & \text{for} \quad a < x_1 < 2a. \end{cases} \tag{1.3.22}$$

and $\psi_{i2} = 0$ for $i = 2, 3, \ldots, N$. From (1.2.7), (1.3.21) and (1.3.22) it follows that

$$2\mathcal{R}\left\{ \left[\sum_\alpha L_{12\alpha}\eta_\alpha \right] g'(x_1) \right\} = \begin{cases} -P_0 & \text{for} \quad -2a < x_1 < -a, \\ P_0 & \text{for} \quad -a < x_1 < a, \\ -P_0 & \text{for} \quad a < x_1 < 2a, \end{cases} \tag{1.3.23}$$

$$2\mathcal{R}\left[\sum_\alpha L_{i2\alpha}\eta_\alpha \right] = 0 \quad \text{for} \quad i = 2, 3, \ldots, N, \tag{1.3.24}$$

where $\mathcal{R}$ denotes the real part of a complex number. Now suppose the η_α are chosen such that

$$\sum_\alpha L_{12\alpha}\eta_\alpha = 1, \qquad \sum_\alpha L_{i2\alpha}\eta_\alpha = 0 \quad \text{for} \quad i = 2, 3, \ldots, N. \tag{1.3.25}$$

Then by Cauchy's theorem and (1.3.23) the appropriate form for $g(z)$ is

$$g'(z) = \frac{1}{\pi i} \int_{-2a}^{-a} \frac{P_0 \, dt}{t-z} - \frac{1}{\pi i} \int_{-a}^{a} \frac{P_0 \, dt}{t-z} + \frac{1}{\pi i} \int_{a}^{2a} \frac{P_0 \, dt}{t-z}$$

$$= \frac{1}{\pi i} \left\{ \log \left(\frac{z+a}{z+2a} \right) - \log \left(\frac{z-a}{z+a} \right) + \log \left(\frac{z-2a}{z-a} \right) \right\}. \tag{1.3.26}$$

Returning to (1.3.25) in view of (1.3.18) it is possible to write solutions to (1.3.25) in the form

$$\eta_\alpha = \omega_\alpha, \qquad \eta_\alpha = \omega_\alpha + \xi_\alpha, \tag{1.3.27}$$

where $\eta_\alpha = \omega_\alpha$ is a particular solution of (1.3.25) and ξ_α is a non-trivial solution to the associated homogeneous system (1.3.18). Hence, from (1.3.21) and (1.3.27) it follows that two possible solutions to the problem are

$$f_\alpha^{(1)}(z_\alpha) = \omega_\alpha g(z_\alpha), \qquad f_\alpha^{(2)}(z_\alpha) = (\omega_\alpha + \xi_\alpha) g(z_\alpha), \tag{1.3.28}$$

where $g(z_\alpha)$ is obtained by integrating (1.3.26). Hence, since the differential equations are linear

$$f_\alpha(z_\alpha) = f_\alpha^{(2)}(z_\alpha) - f_\alpha^{(1)}(z_\alpha)$$

$$= \xi_\alpha g(z_\alpha) \tag{1.3.29}$$

also yields a solution to the system (1.1.1) through (1.2.4). Further it is a solution for which $P_i = 0$ on $x_2 = 0$.

From (1.1.3), (1.2.5) and (1.1.1)

$$\int_R \psi_{ij} \frac{\partial \phi_i}{\partial x_j} \, dR = \int_C \psi_{ij} \phi_i n_j \, ds$$

$$= \int_C P_i \phi_i \, ds \tag{1.3.30}$$

where R is that part of disc $|z| < r$ lying in $x_2 < 0$, and C is the boundary of this region. Now if the ϕ_i and P_i in (1.3.30) are given by (1.2.4), (1.2.6), (1.2.7), (1.3.29) and (1.3.26) then

$$\phi_k' = 2\mathcal{R} \left\{ \sum_\alpha A_{k\alpha} \xi_\alpha \left[\log \frac{z_\alpha + a}{z_\alpha + 2a} - \log \frac{z_\alpha + a}{z_\alpha - a} + \log \frac{z_\alpha - 2a}{z_\alpha - a} \right] \right\}, \tag{1.3.31}$$

$$\psi_{ij} = 2\mathcal{R} \left\{ \sum_\alpha L_{ij\alpha} \xi_\alpha \left[\log \frac{z_\alpha + a}{z_\alpha + 2a} - \log \frac{z_\alpha + a}{z_\alpha - a} + \log \frac{z_\alpha - 2a}{z_\alpha - a} \right] \right\}. \tag{1.3.32}$$

If the radius r of the circular section of the boundary C is now required to tend to infinity then $P_i = 0$ on C so that the right hand side of (1.3.30) is zero. Hence

$$\int_R \psi_{ij} \frac{\partial \phi_i}{\partial x_j} \, dR = 0. \tag{1.3.33}$$

But the ellipticity condition (1.1.3) (in conjunction with (1.2.5)) indicates that (1.3.33) can only hold if the integrand is zero for all points in R. Clearly, from (1.3.31) and (1.3.32) this will not be the case and hence, the assumption that there exists $\xi_\alpha \alpha = 1, 2, \ldots, N$, not all zero, such that (1.3.18) is satisfied, is invalid. Thus the $L_{i2\alpha}$ are linearly independent.

Theorem 1.3.5 *The elements $B_{\alpha\alpha}$ and $C_{\alpha\alpha}$, $\alpha = 1, 2, \ldots, N$ of the matrices $[B_{ij}]$ and $[C_{ij}]$, respectively, have zero real part.*

Proof This result follows readily from Theorem 1.3.2 since, from (1.3.7)

$$\sum_\beta \bar{A}_{i\beta} L_{i2\alpha} \bar{M}_{\beta j} = -\sum_\beta A_{i\alpha} \bar{L}_{i2\beta} \bar{M}_{\beta j}$$

$$= -A_{j\alpha}.$$

Hence

$$\sum_\beta \bar{A}_{i\beta} \bar{M}_{\beta j} \sum_\alpha L_{i2\alpha} M_{\alpha k} = -\sum_\alpha A_{j\alpha} M_{\alpha k}$$

so that

$$B_{jk} + \bar{B}_{kj} = 0,$$

thus showing that $B_{11}, B_{22}, \ldots, B_{NN}$ have zero real part.
  Similarly, from (1.3.7) and (1.2.10)

$$\sum_\alpha \bar{A}_{i\beta} L_{i2\alpha} N_{\alpha j} = -\sum_\alpha A_{i\alpha} \bar{L}_{i2\beta} N_{\alpha j}$$

$$= -L_{j2\beta}.$$

Hence

$$\sum_\beta \bar{A}_{i\beta} \bar{N}_{\beta k} \sum_\alpha L_{i2\alpha} N_{\alpha j} = -\sum_\beta \bar{L}_{j2\beta} \bar{N}_{\beta k}$$

so that

$$C_{kj} + \bar{C}_{jk} = 0,$$

thus showing that $C_{11}, C_{22}, \ldots, C_{NN}$ have zero real part.

Theorem 1.3.6 *The matrices $[B_{kj}]$ and $[C_{kj}]$ are non-singular.*

Proof Both of these matrices are the products of two non-singular matrices and hence the result follows trivially from the product rule for determinants.

Theorem 1.3.7 *The matrix* $(B_{kj} - \bar{B}_{kj})$ *is non-singular.*

Proof To establish this result consider particular analytic functions $f_\alpha(z)$ of the form

$$f_\alpha(z) = \frac{1}{4\pi} M_{\alpha j} d_j \log z, \tag{1.3.34}$$

where the d_j are real constants. Hence in (1.2.4) and (1.2.7)

$$\phi_k = \frac{1}{4\pi} \left\{ \sum_\alpha A_{k\alpha} M_{\alpha j} \log z_\alpha + \sum_\alpha \bar{A}_{k\alpha} \bar{M}_{\alpha j} \log \bar{z}_\alpha \right\} d_j, \tag{1.3.35}$$

$$\psi_{ij} = \frac{1}{4\pi} \left\{ \sum_\alpha L_{ij\alpha} M_{\alpha j} z_\alpha^{-1} + \sum_\alpha \bar{L}_{ij\alpha} \bar{M}_{\alpha j} \bar{z}_\alpha^{-1} \right\} d_j. \tag{1.3.36}$$

The ϕ_k in (1.3.35) is a solution to (1.1.1) valid in a region S in the plane cut from the origin along the positive real axis.

Let

$$E = \int_S a_{ijkl} \frac{\partial \phi_i}{\partial x_j} \frac{\partial \phi_k}{\partial x_l} \, dS = \int_C \psi_{ij} \phi_i n_j \, dS \tag{1.3.37}$$

where C is the boundary of the region S and n_j is the outward normal to C. In particular let S consist of the region bounded by the circles $|z| = r$ and $|z| = R$. The region is made simply connected by making a cut in the plane from $x_1 = r$ to $x_1 = R$. (see Fig. 1.3.1). The line integral (1.3.37) is now evaluated for this particular contour.

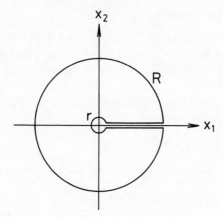

Fig. 1.3.1

Let $z = x_1 + ix_2 = R \exp(i\theta)$ so that

$$z_\alpha = x_1 + \tau_\alpha x_2 = R(\cos\theta + \tau_\alpha \sin\theta). \tag{1.3.38}$$

Since $ds = R\, d\theta$ it is apparent from (1.3.36) and (1.3.38) that $\psi_{ij}\, ds$ is independent of R. Also (1.3.35) adopts the form

$$\phi_k = \xi_k \log R + g_k(\theta)$$

where the constants ξ_k and $g(\theta)$ can be determined by substituting (1.3.38) in (1.3.35). However their precise form is unimportant here. Hence the part of the line integral round $|z| = R$ involving R is

$$\xi_k \log R \int_0^{2\pi} \psi_{ij} n_j R\, d\theta = 0,$$

the integral being zero since ψ_{ij} is single-valued. Hence the integral round the circle $|z| = R$ is independent of the radius R. Hence the integral round the smaller circle $|z| = r$ will have the same magnitude as the integral round $|z| = R$ but be opposite in sign. Thus the two will cancel out so that the line integral (1.3.37) reduces to the integral over the cut. That is

$$E = \int_r^R \psi_{i2}(x_1, 0)[\phi_i(x_1, 0-) - \phi_i(x_1, 0+)]\, dx_1$$

$$= (2\pi)^{-1} i(B_{ij} - \bar{B}_{ij})\, d_i\, d_j \log(R/r). \tag{1.3.39}$$

From (1.1.3) and (1.3.37) it is apparent that E must be positive for all non-zero values of d_i and hence it follows immediately from (1.3.39) that the matrix $i(B_{ij} - \bar{B}_{ij})$ must be positive definite. Hence $[B_{ij} - \bar{B}_{ij}]$ is a non-singular matrix.

1.4 Uniqueness of solution

The latter part of the proof of Theorem 1.3.4 (from (1.3.28) onwards) is essentially concerned with the uniqueness of the solution for the particular boundary-value problem considered in the theorem. The same procedure may readily be used to investigate the uniqueness of the solution to the general boundary-value problem of Section 1.1. For suppose that it is possible to obtain two solutions

$$\phi_k^{(1)},\ \psi_{ij}^{(1)} \quad \text{and} \quad \phi_k^{(2)},\ \psi_{ij}^{(2)}$$

for a problem in which ϕ_k is prescribed on C_1, P_i on C_2 with $C = C_1 + C_2$. Then

$$\phi_k = \phi_k^{(1)} - \phi_k^{(2)}, \tag{1.4.1}$$

$$\psi_{ij} = \psi_{ij}^{(1)} - \psi_{ij}^{(2)} \tag{1.4.2}$$

is also a solution to (1.1.1). Furthermore $\phi_k = 0$ on C_1 and $P_i = 0$ on C_2. Hence

$$\int_R \psi_{ij} \frac{\partial \phi_i}{\partial x_j} \, dR = \int_C P_i \phi_i \, ds = 0. \tag{1.4.3}$$

From the ellipticity condition (1.1.3) and (1.2.5) it immediately follows that (1.4.3) can only hold if all the $\partial \phi_i / \partial x_j$, are zero. Since the ψ_{ij} are just linear combinations of these derivatives it follows that $\psi_{ij} = 0$ and hence, from (1.4.2) $\psi_{ij}^{(1)} = \psi_{ij}^{(2)}$. Also, from (1.4.1) it follows that $\phi_k^{(1)}$ and $\phi_k^{(2)}$ can only differ by a constant. If ϕ_k is prescribed on at least part of the boundary C then the ϕ_k will be uniquely determined.

2

Physical problems governed by the system

2.1 Introduction

In this chapter a number of particular classes of physical problems which are governed by special cases of the system (1.1.1) are presented. Where possible, explicit forms for the constants occurring in Chapter 1 will be obtained. When this is not possible, further properties of these constants will be derived for the particular class of physical problems under consideration.

2.2 Laplace's equation

The simplest special case of (1.1.1) is Laplace's equation in two spatial variables

$$\frac{\partial^2 \phi_1}{\partial x_1^2} + \frac{\partial^2 \phi_1}{\partial x_2^2} = 0. \tag{2.2.1}$$

This equation is the governing equation for a large number of physical problems and it is inappropriate to attempt to detail all of these problems here. However, it is appropriate to note that all the analysis of subsequent sections applies to Laplace's equation and, further, to derive the explicit forms for the constants of Chapter 1 which are relevant to (2.2.1). To obtain these constants it is first observed that (2.2.1) may be obtained from (1.1.1) by putting $N = 1$, $a_{1111} = a_{1212} = K > 0$ and all the other a_{ijkl} zero. Hence

$$P_1 = K \left[\frac{\partial \phi_1}{\partial x_1} n_1 + \frac{\partial \phi_1}{\partial x_2} n_2 \right]$$

$$= K \frac{\partial \phi_1}{\partial n}. \tag{2.2.2}$$

Use of the equations of Section 1.2 shows that, in this case, all the constants are zero except for those listed below (together with their conjugates)

$$\tau_1 = i, \quad A_{11} = 1, \quad L_{111} = K, \quad L_{121} = iK, \quad N_{11} = 1,$$
$$M_{11} = -iK^{-1}, \quad C_{11} = iK, \quad B_{11} = -iK^{-1}. \tag{2.2.3}$$

The physical significance of the constant K in (2.2.2) will, of course, depend on the particular physical system in question. For future reference, two different physical systems governed by Laplace's equation will now be examined.

(A) Antiplane deformations of isotropic elastic materials

Here the antiplane stresses σ_{13}, σ_{23} are related to the antiplane displacement u_3 by the equations

$$\sigma_{13} = \mu \frac{\partial u_3}{\partial x_1}, \tag{2.2.4}$$

$$\sigma_{23} = \mu \frac{\partial u_3}{\partial x_2}, \tag{2.2.5}$$

where μ is the shear modulus. The equation of elastic equilibrium is

$$\frac{\partial \sigma_{13}}{\partial x_1} + \frac{\partial \sigma_{23}}{\partial x_2} = 0. \tag{2.2.6}$$

Substitution of (2.2.4) and (2.2.5) into (2.2.6) yields

$$\mu \frac{\partial^2 u_3}{\partial x_1^2} + \mu \frac{\partial u_3}{\partial x_2^2} = 0, \tag{2.2.7}$$

so that the system (1.1.1) is applicable with $N = 1$, $\phi_1 \equiv u_3$, $a_{1111} = a_{1212} = \mu$, $\psi_{11} \equiv \sigma_{13}$ and $\psi_{12} \equiv \sigma_{23}$. Hence the K in (2.2.2) and (2.2.3) is, in this case, equal to the shear modulus μ and

$$P_1 = \mu \left(\frac{\partial u_3}{\partial x_1} n_1 + \frac{\partial u_3}{\partial x_2} n_2 \right) = \sigma_{13} n_1 + \sigma_{23} n_2 \tag{2.2.8}$$

is just the traction at a point P on a surface through P with outward normal $\mathbf{n}$.

(B) Plane thermostatics for an isotropic material

For a static temperature field $T(x_1, x_2)$ in an isotropic material the heat fluxes across two planes parallel to the coordinate axes are

$$f_1 = -\lambda \frac{\partial T}{\partial x_1},$$

(2.2.9)

$$f_2 = -\lambda \frac{\partial T}{\partial x_2},$$

(2.2.10)

where λ is the thermal conductivity of the substance.

By considering the flux across an element of volume of the solid it may readily be shown that (Carslaw and Jaeger [5]).

$$-\frac{\partial f_1}{\partial x_1} - \frac{\partial f_2}{\partial x_2} = 0.$$

(2.2.11)

Substitution of (2.2.9) and (2.2.10) into (2.2.11) yields

$$\lambda \frac{\partial^2 T}{\partial x_1^2} + \lambda \frac{\partial^2 T}{\partial x_2} = 0$$

(2.2.12)

so that the system (1.1.1) is applicable with $N = 1$, $\phi_1 \equiv T$, $a_{1111} = a_{1212} = \lambda$, $\psi_{11} \equiv -f_1$ and $\psi_{12} \equiv -f_2$. Hence the K in (2.2.2) and (2.2.3) is, in this case, the thermal conductivity λ and

$$P_1 = \lambda \left(\frac{\partial T}{\partial x_1} n_1 + \frac{\partial T}{\partial x_2} n_2 \right) = \lambda \frac{\partial T}{\partial n}.$$

(2.2.13)

Here, $-P_1$ is the flux of heat at a point across the surface with outward normal $\mathbf{n}$.

2.3 Generalized plane thermostatics for anisotropic materials

For a static temperature field $T(x_1, x_2)$ in an anisotropic material the heat flux across planes parallel to the coordinate axes is

$$-f_i = \lambda_{i1} \frac{\partial T}{\partial x_1} + \lambda_{i2} \frac{\partial T}{\partial x_2} \quad \text{for} \quad i = 1, 2, 3,$$

(2.3.1)

where the λ_{ij} are the conductivity coefficients which satisfy the symmetry property $\lambda_{ij} = \lambda_{ji}$. Equation (2.2.11) is valid for this case so use of (2.2.1) in (2.2.11) yields

$$\lambda_{ij} \frac{\partial^2 T}{\partial x_i \, \partial x_j} = 0.$$

(2.3.2)

Note that even though the temperature T is independent of x_3 the flux $-f_3$ need not be zero and in this sense a problem governed by (2.3.1) and (2.3.2) is termed a generalized plane problem. If λ_{31} and λ_{32} are both zero then the flux across the planes $x_3 = $ constant will be zero and then

(2.3.1) and (2.3.2) are the governing equations for the temperature and flux in plane thermostatics.

It is apparent that the system (1.1.1) is applicable with $N = 1$, $\phi_1 \equiv T$, $a_{1i1j} = \lambda_{ij}$ for $i, j = 1, 2, 3$.

It is worth noting at this point that here the a_{ijkl} have been defined for $j, l = 1, 2, 3$ whereas (1.1.1) only calls for them to be defined for $j, l = 1, 2$. This presents no difficulty since obviously the equation (1.1.1) is not affected nor is the general boundary value problem posed in Section 1.1. However such an extension does permit the ψ_{ij} in (1.2.5) and $L_{ij\alpha}$ in (1.2.8) to be defined for more values of the subscript j if this is desirable.

Hence, from (1.2.5) and (2.2.1)

$$\psi_{1j} = -f_j \quad \text{for} \quad j = 1, 2, 3 \tag{2.3.3}$$

and the flux $-P_1$ at a point across the surface with outward normal $\mathbf{n}$ is given through

$$P_1 = a_{1j1l} \frac{\partial T}{\partial x_l} n_j = \lambda_{ij} \frac{\partial T}{\partial x_j} n_i$$

$$= \lambda_{1j} \frac{\partial T}{\partial x_j} n_1 + \lambda_{2j} \frac{\partial T}{\partial x_j} n_2 + \lambda_{3j} \frac{\partial T}{\partial x_j} n_3. \tag{2.3.4}$$

Of course, for a generalized plane problem, the normal on the boundary is of the form $\mathbf{n} = (n_1, n_2, 0)$ so (2.3.4) reduces to

$$P_1 = \lambda_{1j} \frac{\partial T}{\partial x_j} n_1 + \lambda_{2j} \frac{\partial T}{\partial x_j} n_2. \tag{2.3.5}$$

From the equations of Section 1.2 it follows that the non-zero constants for generalized plane thermostatics for anisotropic materials take the form

$$\tau_1 = \frac{-\lambda_{12} + i(\lambda_{11}\lambda_{22} - \lambda_{12}^2)^{1/2}}{\lambda_{22}}, \tag{2.3.6}$$

$$\left.\begin{array}{l} A_{11} = 1, \qquad N_{11} = 1, \qquad L_{1i1} = \lambda_{i1} + \tau_1 \lambda_{i2}, \qquad i = 1, 2, 3, \\[2mm] M_{11} = L_{121}^{-1} = (\lambda_{21} + \tau_1 \lambda_{22})^{-1}, \qquad C_{11} = \lambda_{21} + \tau_1 \lambda_{22}, \\[2mm] B_{11} = (\lambda_{21} + \tau_1 \lambda_{22})^{-1}. \end{array}\right\} \tag{2.3.7}$$

2.4   Generalized plane deformations of anisotropic elastic materials.

The stresses σ_{ij} are related to the elastic displacements u_k by the equations

$$\sigma_{ij} = c_{ijkl} \frac{\partial u_k}{\partial x_l}, \tag{2.4.1}$$

where $i, j, k, l = 1, 2, 3$ and the elastic constants c_{ijkl} have the symmetry properties

$$c_{ijkl} = c_{jikl} = c_{ijlk} = c_{klij}.$$ (2.4.2)

The equilibrium equations are

$$\frac{\partial \sigma_{ij}}{\partial x_j} = 0.$$ (2.4.3)

Substitutions of (2.4.1) into (2.4.3) yields

$$c_{ijkl} \frac{\partial^2 u_k}{\partial x_j \, \partial x_l} = 0.$$ (2.4.4)

The strain energy density must be everywhere positive and hence

$$c_{ijkl} \frac{\partial u_k}{\partial x_l} \frac{\partial u_i}{\partial x_j} > 0.$$ (2.4.5)

Now suppose the displacement u_k is independent of x_3. Then (1.1.1) is applicable with $N = 3$,

$$\phi_i \equiv u_i \quad \text{for} \quad i = 1, 2, 3$$ (2.4.6)

$$\sigma_{ij} \equiv \psi_{ij} \quad \text{for} \quad i = 1, 2, 3 \text{ and } j = 1, 2$$ (2.4.7)

$$a_{ijkl} = c_{ijkl} \quad \text{for} \quad i, k = 1, 2, 3 \text{ and } j, l = 1, 2.$$ (2.4.8)

Also, the condition (2.4.5) that the strain energy density be everywhere positive simply becomes the ellipticity condition (1.1.3). The P_i in (1.1.4) becomes

$$P_i = c_{ijkl} \frac{\partial u_k}{\partial x_l} n_j \quad \text{for} \quad i = 1, 2, 3$$ (2.4.9)

and this is the traction vector at a point on a surface with normal n_j. The definitions of the constants occurring in Section 1.2 are all applicable to the present case with the sum over α going from one to three. Hence the $A_{k\alpha}$ are the non-trivial solutions of

$$(c_{ijkl} + c_{i1k2}\tau_\alpha + c_{i2k1}\tau_\alpha + c_{i2k2}\tau_\alpha^2)A_{k\alpha} = 0 \quad \text{for} \quad \alpha = 1, 2, 3$$ (2.4.10)

where the τ_α are the three roots with positive real part of

$$|c_{i1k1} + c_{i1k2}\tau + c_{i2k1}\tau + c_{i2k2}\tau^2| = 0.$$ (2.4.11)

The $L_{ij\alpha}$ are given by

$$L_{ij\alpha} = (c_{ijk1} + \tau_\alpha c_{ijk2})A_{k\alpha}$$ (2.4.12)

while $[N_{\alpha j}]$ and $[M_{\alpha j}]$ are the 3×3 inverses of $[A_{k\alpha}]$ and $[L_{i2\alpha}]$ respectively. Finally,

$$C_{ik} = \sum_{\alpha} L_{i2\alpha} N_{\alpha k}, \tag{2.4.13}$$

$$B_{ij} = \sum_{\alpha} A_{k\alpha} M_{\alpha j}, \tag{2.4.14}$$

where the sum over α is from one to three.

As in the previous section, the a_{ijkl} are defined, through (2.4.8), for values of j greater than two. This causes no difficulty in applying the system of Section 1.1. However, it does not permit P_1 in (2.4.9) to be defined for a normal with a non-zero n_3 component (although on any boundary of the material n_3 must, of necessity, be zero for a generalized plane problem). Also, the extension of the definition of a_{ijkl} permits the stresses σ_{13}, σ_{23} and σ_{33} to be defined. In the case of σ_{13} and σ_{23} this is unnecessary since, by the symmetry properties of the c_{ijkl}, $\sigma_{31} = \sigma_{13}$ and $\sigma_{32} = \sigma_{23}$ so attention may be restricted to σ_{31} and σ_{32}.

2.5   Properties of the constants occurring in the elastic equations

The aim in this section is to determine the form of the constants in Section 2.4 when the $x_i = 0$, $i = 1, 2, 3$ plane is a plane of elastic symmetry. The analysis closely follows that of Clements [10].

Elastic symmetry with respect to the plane $x_1 = 0$.   When such symmetry exists the elastic constants c_{ijkl} with an uneven number of ones occurring in their subscripts are zero so that the equations (2.4.10) become

$$\begin{bmatrix} c_{1111} + \tau^2 c_{1212} & (c_{1122} + c_{1221})\tau & (c_{1132} + c_{1231})\tau \\ (c_{2112} + c_{2211})\tau & c_{2121} + \tau^2 c_{2222} & c_{2131} + \tau^2 c_{2232} \\ (c_{3112} + c_{3211})\tau & c_{3121} + \tau^2 c_{3222} & c_{3131} + \tau^2 c_{3232} \end{bmatrix} \begin{bmatrix} A_1 \\ A_2 \\ A_3 \end{bmatrix} = 0 \tag{2.5.1}$$

where the values of τ are obtained by solving equation (2.4.11) which now reduces to a cubic in τ^2. Two possible cases need to be considered.

Case 1.   Suppose the roots of the cubic are real. Then since real values of τ have been excluded, the cubic in τ^2 must have negative roots. Hence the six values of τ are complex numbers with zero real part. Then if the $A_{3\alpha}$ are put equal to one, equation (2.5.1) has a solution for which the $A_{1\alpha}$ have zero real part and the $A_{2\alpha}$ have zero imaginary part. In order to determine the nature of the $L_{ij\alpha}$ consider equation (2.4.12). Written out in full it is

$$L_{ij\alpha} = (c_{ij11} + \tau_\alpha c_{ij12})A_{1\alpha} + (c_{ij21} + \tau_\alpha c_{ij22})A_{2\alpha} + (c_{ij31} + \tau_\alpha c_{ij32})A_{3\alpha}.$$
(2.5.2)

Suppose both or neither of the subscripts i, j are equal to one, then

$$L_{ij\alpha} = c_{ij11}A_1 + \tau_\alpha c_{ij22}A_{2\alpha} + \tau_\alpha c_{ij32}A_{3\alpha}$$
(2.5.3)

and remembering the properties of the $A_{k\alpha}$ and τ_α this clearly has zero real part. Suppose only one of the subscripts i, j is equal to one, then

$$L_{ij\alpha} = \tau_\alpha c_{ij12}A_{1\alpha} + c_{ij21}A_{2\alpha} + c_{ij31}A_{3\alpha}$$
(2.5.4)

and in this case the $L_{ij\alpha}$ have zero imaginary part. Hence in matrix form

$$[A_{i\alpha}] = \begin{bmatrix} ia''_{11} & ia''_{12} & ia''_{13} \\ a'_{21} & a'_{22} & a'_{23} \\ 1 & 1 & 1 \end{bmatrix},$$
(2.5.5)

$$[L_{i2\alpha}] = \begin{bmatrix} l'_{11} & l'_{12} & l'_{13} \\ il''_{21} & il''_{22} & il''_{23} \\ il''_{31} & il''_{32} & il''_{33} \end{bmatrix}$$
(2.5.6)

where $A_{i\alpha} = a'_{i\alpha} + ia''_{i\alpha}$ and $L_{i2\alpha} = l'_{i\alpha} + il''_{i\alpha}$ with $a'_{i\alpha}$, $a''_{i\alpha}$, $l'_{i\alpha}$ and $l''_{i\alpha}$ real. Hence the inverses $[N_{\alpha j}]$ and $[M_{\alpha j}]$ of these two matrices adopt the forms

$$[N_{\alpha j}] = \begin{bmatrix} in''_{11} & n'_{12} & n'_{13} \\ in''_{21} & n'_{22} & n'_{23} \\ in''_{31} & n'_{32} & n'_{33} \end{bmatrix},$$
(2.5.7)

$$[M_{\alpha j}] = \begin{bmatrix} m'_{11} & im''_{12} & im''_{13} \\ m'_{21} & im''_{22} & im''_{23} \\ m'_{22} & im''_{32} & im''_{33} \end{bmatrix},$$
(2.5.8)

where $N_{\alpha j} = n'_{\alpha j} + in''_{\alpha j}$ and $M_{\alpha j} = m'_{\alpha j} + im''_{\alpha j}$ with $n'_{\alpha j}$, $n''_{\alpha j}$, $m'_{\alpha j}$ and $m''_{\alpha j}$ real. Equations (2.4.13) and (2.4.14) may now be used in conjunction with (2.5.5)–(2.5.8) to yield the following forms for $[C_{ik}]$ and $[B_{ij}]$

$$[C_{ik}] = \begin{bmatrix} ic''_{11} & c'_{12} & c'_{13} \\ c'_{21} & ic''_{22} & ic''_{23} \\ c'_{31} & ic''_{32} & ic''_{33} \end{bmatrix},$$
(2.5.9)

$$[B_{ij}] = \begin{bmatrix} ib''_{11} & b'_{12} & b'_{13} \\ b'_{21} & ib''_{22} & ib''_{23} \\ b'_{31} & ib''_{32} & ib''_{33} \end{bmatrix}$$
(2.5.10)

where $C_{ik} = c'_{ik} + c''_{ik}$ and $B_{ij} = b'_{ij} + ib''_{ij}$ where, as before, the lower case subscripted constants are real.

Case 2. Suppose only one of the roots (τ_1, say) of the cubic is real and negative. Then the other roots form a conjugate pair and $\tau_2 = -\bar{\tau}_3$. Putting the $A_{3\alpha}$ equal to one equation (2.5.1) has a solution for which A_{11} has zero real part and A_{21} has zero imaginary part. Also $A_{12} = -\bar{A}_{13}$ and $A_{22} = \bar{A}_{23}$. Suppose both or neither of the subscripts i, j of $L_{ij\alpha}$ are equal to one, then using (2.5.3) it follows that the L_{ij1} have zero real part and $L_{ij2} = -\bar{L}_{ij3}$. For the case when only one of the subscripts i, j of $L_{ij\alpha}$ is equal to one it follows from (2.5.4) that the L_{ij1} have zero imaginary part and $L_{ij2} = \bar{L}_{ij3}$. In matrix form

$$[A_{i\alpha}] = \begin{bmatrix} ia''_{11} & A_{12} & -\bar{A}_{12} \\ a'_{21} & A_{22} & \bar{A}_{22} \\ 1 & 1 & 1 \end{bmatrix}, \tag{2.5.11}$$

$$[L_{i2\alpha}] = \begin{bmatrix} l'_{11} & L_{122} & \bar{L}_{122} \\ il''_{21} & L_{222} & -\bar{L}_{222} \\ il''_{31} & L_{322} & -\bar{L}_{322} \end{bmatrix}. \tag{2.5.12}$$

Hence

$$[N_{\alpha j}] = \begin{bmatrix} in''_{11} & n'_{12} & n'_{13} \\ N_{21} & N_{22} & N_{23} \\ -\bar{N}_{21} & \bar{N}_{22} & \bar{N}_{23} \end{bmatrix}, \tag{2.5.13}$$

$$[M_{\alpha j}] = \begin{bmatrix} m'_{11} & im''_{12} & im''_{13} \\ M_{21} & M_{22} & M_{23} \\ \bar{M}_{21} & -\bar{M}_{22} & -\bar{M}_{23} \end{bmatrix}, \tag{2.5.14}$$

$$[C_{ik}] = \begin{bmatrix} ic''_{11} & c'_{12} & c'_{13} \\ c'_{21} & ic''_{22} & ic''_{23} \\ c'_{31} & ic''_{32} & ic''_{33} \end{bmatrix}, \tag{2.5.15}$$

$$[B_{ij}] = \begin{bmatrix} ib''_{11} & b'_{12} & b'_{13} \\ b'_{21} & ib''_{22} & ib''_{23} \\ b'_{31} & ib''_{32} & ib''_{33} \end{bmatrix}. \tag{2.5.16}$$

Elastic symmetry with respect to the plane $x_2 = 0$. In this case the elastic constants c_{ijkl} with an uneven number of twos occurring in their subscripts are zero so that (2.4.10) becomes

$$\begin{bmatrix} c_{1111} + \tau^2 c_{1212} & (c_{1122} + c_{1221})\tau & c_{1131} + \tau^2 c_{1232} \\ (c_{1122} + c_{1221})\tau & c_{2121} + \tau^2 c_{2222} & (c_{2231} + c_{2132})\tau \\ c_{1131} + \tau^2 c_{3212} & (c_{3122} + c_{3221})\tau & c_{3131} + \tau^2 c_{3232} \end{bmatrix} \begin{bmatrix} A_1 \\ A_2 \\ A_3 \end{bmatrix} = 0 \quad (2.5.17)$$

where the values of τ are given by (2.4.11) which reduces to a cubic in τ^2. As in the case of elastic symmetry with respect to the plane $x_1 = 0$, it is necessary to consider two possible cases.

Case 1. Suppose the roots of the cubic are real and negative. Then the six values of τ are complex with zero real part. Hence, if the $A_{3\alpha}$ are put equal to one equation (2.5.17) has a solution for which the $A_{1\alpha}$ have zero imaginary part and the $A_{2\alpha}$ have zero real part. Also, using (2.4.12) it may readily be shown that with the choice of the $A_{i\alpha}$ the $L_{12\alpha}$ and $L_{32\alpha}$ have zero real part and the $L_{22\alpha}$ have zero imaginary part. In matrix form

$$[A_{i\alpha}] = \begin{bmatrix} a'_{11} & a'_{12} & a'_{13} \\ ia''_{21} & ia''_{22} & ia''_{23} \\ 1 & 1 & 1 \end{bmatrix}, \tag{2.5.18}$$

$$[L_{i2\alpha}] = \begin{bmatrix} il''_{11} & il''_{12} & il''_{13} \\ l'_{21} & l'_{22} & l'_{23} \\ il''_{31} & il''_{32} & il''_{33} \end{bmatrix}. \tag{2.5.19}$$

Hence

$$[N_{\alpha j}] = \begin{bmatrix} n'_{11} & in''_{12} & n'_{13} \\ n'_{21} & in''_{22} & n'_{23} \\ n'_{31} & in''_{32} & n'_{33} \end{bmatrix}, \tag{2.5.20}$$

$$[M_{\alpha j}] = \begin{bmatrix} im''_{11} & m'_{12} & im''_{13} \\ im''_{21} & m'_{22} & im''_{23} \\ im''_{31} & m'_{32} & im''_{33} \end{bmatrix}, \tag{2.5.21}$$

$$[C_{ik}] = \begin{bmatrix} ic''_{11} & c''_{12} & ic''_{13} \\ c'_{21} & ic''_{22} & c'_{23} \\ ic''_{31} & c'_{32} & ic''_{33} \end{bmatrix}, \tag{2.5.22}$$

$$[B_{ij}] = \begin{bmatrix} ib''_{11} & b'_{12} & ib''_{13} \\ b'_{21} & ib''_{22} & b'_{23} \\ ib''_{31} & b'_{32} & ib''_{33} \end{bmatrix}. \tag{2.5.23}$$

Case 2. Suppose only one of the roots (τ_1, say) of the cubic is real and negative. Then the other roots form a conjugate pair and $\tau_2 = -\bar{\tau}_3$. Putting the $A_{3\alpha}$ equal to one equation (2.5.17) has a solution for which A_{11} has zero imaginary part and A_{21} has zero real part. Also $A_{22} = -\bar{A}_{23}$ and $A_{12} = \bar{A}_{13}$. Then using (2.4.12) it may easily be shown that L_{121} and L_{321} have zero real part and that $L_{122} = -\bar{L}_{123}$ and $L_{322} = -\bar{L}_{323}$. Also

L_{221} has zero imaginary part and $L_{222} = -\bar{L}_{223}$. In matrix form

$$[A_{i\alpha}] = \begin{bmatrix} a'_{11} & A_{12} & \bar{A}_{12} \\ ia''_{21} & A_{22} & -\bar{A}_{22} \\ 1 & 1 & 1 \end{bmatrix}, \tag{2.5.24}$$

$$[L_{i2\alpha}] = \begin{bmatrix} il''_{11} & L_{122} & -\bar{L}_{122} \\ l'_{12} & L_{222} & \bar{L}_{222} \\ il''_{31} & L_{232} & -\bar{L}_{232} \end{bmatrix}. \tag{2.5.25}$$

Hence

$$[N_{\alpha j}] = \begin{bmatrix} n'_{11} & in''_{12} & n_{13} \\ N_{21} & N_{22} & N_{23} \\ \bar{N}_{21} & -\bar{N}_{22} & \bar{N}_{23} \end{bmatrix}, \tag{2.5.26}$$

$$[M_{\alpha j}] = \begin{bmatrix} im''_{11} & m'_{12} & im''_{13} \\ M_{21} & M_{22} & M_{23} \\ -\bar{M}_{21} & \bar{M}_{22} & -\bar{M}_{23} \end{bmatrix}, \tag{2.5.27}$$

$$[C_{ik}] = \begin{bmatrix} ic''_{11} & c'_{12} & ic''_{13} \\ c'_{21} & ic''_{22} & c'_{23} \\ ic''_{31} & c'_{32} & ic''_{33} \end{bmatrix}, \tag{2.5.28}$$

$$[B_{ij}] = \begin{bmatrix} ib''_{11} & b'_{12} & ib''_{13} \\ b'_{21} & ib''_{22} & b'_{23} \\ ib''_{31} & b'_{32} & ib''_{33} \end{bmatrix}. \tag{2.5.29}$$

Elastic symmetry with respect to the plane $x_3 = 0$. In this case the elastic constants c_{ijkl} with an odd number of threes in their subscripts are zero. It follows that two of the roots of (2.4.11) are the roots of the quadratic

$$c_{3131} + \tau c_{3132} + \tau c_{3231} + \tau^2 c_{3232} = 0. \tag{2.5.30}$$

Without any loss of generality, the roots of this quadratic may be taken to be τ_1 and $\bar{\tau}_1$. It then follows from (2.4.10) and (2.4.12) that $A_{11} = A_{21} = A_{32} = A_{33} = 0$, $L_{121} = L_{221} = L_{322} = L_{323} = 0$. In matrix form

$$[A_{ij}] = \begin{bmatrix} 0 & A_{12} & A_{13} \\ 0 & A_{22} & A_{23} \\ A_{31} & 0 & 0 \end{bmatrix}, \tag{2.5.31}$$

$$[L_{i2\alpha}] = \begin{bmatrix} 0 & L_{122} & L_{123} \\ 0 & L_{222} & L_{223} \\ L_{321} & 0 & 0 \end{bmatrix}. \tag{2.5.32}$$

Hence

$$[N_{\alpha j}] = \begin{bmatrix} 0 & 0 & N_{13} \\ N_{21} & N_{22} & 0 \\ N_{31} & N_{32} & 0 \end{bmatrix}, \tag{2.5.33}$$

$$[M_{\alpha j}] = \begin{bmatrix} 0 & 0 & M_{13} \\ M_{21} & M_{22} & 0 \\ M_{31} & M_{32} & 0 \end{bmatrix}, \tag{2.5.34}$$

$$[C_{ik}] = \begin{bmatrix} C_{11} & C_{12} & 0 \\ C_{21} & C_{22} & 0 \\ 0 & 0 & C_{33} \end{bmatrix}, \tag{2.5.35}$$

$$[B_{ij}] = \begin{bmatrix} B_{11} & B_{12} & 0 \\ B_{21} & B_{22} & 0 \\ 0 & 0 & B_{33} \end{bmatrix}. \tag{2.5.36}$$

It is of interest to note that when symmetry with respect to the plane $x_3 = 0$ exists the problem uncouples into a plane problem and antiplane problem. This is clear from (2.4.1) which, with u_k independent of x_3, becomes

$$\left. \begin{aligned} \sigma_{11} &= c_{1111} \frac{\partial u_1}{\partial x_1} + c_{1112} \frac{\partial u_1}{\partial x_2} + c_{1121} \frac{\partial u_2}{\partial x_1} + c_{1122} \frac{\partial u_2}{\partial x_2}, \\ \sigma_{12} &= c_{1211} \frac{\partial u_1}{\partial x_1} + c_{1212} \frac{\partial u_1}{\partial x_2} + c_{1221} \frac{\partial u_2}{\partial x_1} + c_{1222} \frac{\partial u_2}{\partial x_2}, \\ \sigma_{22} &= c_{2211} \frac{\partial u_1}{\partial x_1} + c_{2212} \frac{\partial u_1}{\partial x_2} + c_{2221} \frac{\partial u_2}{\partial x_1} + c_{2222} \frac{\partial u_2}{\partial x_2}, \\ \sigma_{33} &= c_{3311} \frac{\partial u_1}{\partial x_1} + c_{3312} \frac{\partial u_1}{\partial x_2} + c_{3321} \frac{\partial u_2}{\partial x_1} + c_{3322} \frac{\partial u_2}{\partial x_2}, \end{aligned} \right\} \tag{2.5.37}$$

$$\left. \begin{aligned} \sigma_{13} &= c_{1331} \frac{\partial u_3}{\partial x_1} + c_{1332} \frac{\partial u_3}{\partial x_2}, \\ \sigma_{23} &= c_{2331} \frac{\partial u_3}{\partial x_1} + c_{2332} \frac{\partial u_3}{\partial x_2}. \end{aligned} \right\} \tag{2.5.38}$$

The plane problem involves the displacements of u_1 and u_2 and the stresses σ_{11}, σ_{12}, σ_{22} and σ_{33}. The antiplane problem involves the displacement u_3 and the stresses σ_{13} and σ_{23}. Substitution of (2.5.37) into (2.4.3) makes apparent the fact that the plane problem is governed by a system of two second order differential equations with the dependent variables being u_1 and u_2. Hence the system (1.1.1) is applicable with $N = 2$. Also, from (2.5.38) and (2.4.3) it is apparent that the antiplane problem is governed by a second order equation and hence (1.1.1) is applicable with $N = 1$. In fact, the governing equation for the antiplane problem is similar in form to the equation (2.2.2) for plane anisotropic thermostatics. Although it is not particularly instructive at this point to pursue the plane and antiplane deformations as entirely separate problems it is worth noting that it is possible to obtain simple expressions for those elements of the matrices (2.5.31)–(2.5.36) which correspond to the antiplane deformation. These expressions are

$$A_{31} = 1, \qquad L_{321} = c_{3231} + \tau_1 c_{3232}, \qquad N_{13} = 1,$$
$$M_{13} = (c_{3231} + \tau_1 c_{3232})^{-1}, \qquad C_{33} = L_{321}, \qquad B_{33} = M_{13} \qquad (2.5.39)$$

Elastic symmetry with respect to two of the planes $x_i = 0$, $i = 1, 2, 3$. In this case all the c_{ijkl} with an odd number of ones, twos and threes in their indices are zero so (2.4.10) becomes

$$\begin{bmatrix} c_{1111} + \tau^2 c_{1212} & (c_{1122} + c_{1221})\tau & 0 \\ (c_{1122} + c_{1221})\tau & c_{2121} + \tau^2 c_{2222} & 0 \\ 0 & 0 & c_{3131} + \tau^2 c_{3232} \end{bmatrix} \begin{bmatrix} A_1 \\ A_2 \\ A_3 \end{bmatrix} = 0, \quad (2.5.40)$$

while (2.4.11) becomes

$$\{c_{1212}c_{2222}\tau^4 + [c_{1111}c_{2222} + c_{1212}^2 - (c_{1122} + c_{1221})^2]\tau^2 + c_{1111}c_{1212}\}$$
$$\times \{c_{3131} + \tau^2 c_{3232}\} = 0. \quad (2.5.41)$$

In this case it is possible to write down explicit expressions for the matrices $[A_{i\alpha}]$, $[L_{i2\alpha}]$ etc. However the expressions for the elements of some of the matrices are rather long and involved and will not be presented here. Rather the explicit forms for the particular case of a transversely isotropic material will be given in a later section.

2.6 Quasi-static deformations of anisotropic elastic materials

The linear equations of motion for a homogeneous anisotropic elastic material are

$$c_{ijkl} \frac{\partial^2 u_k}{\partial x_l \partial x_j} = \rho \frac{\partial^2 u_i}{\partial t^2}, \qquad (2.6.1)$$

where ρ is the density and t denotes time. If the variation in displacement is independent of x_3 and furthermore, depends on x_1, x_2 and t in such a way that it can be written as a function of $x_1 + Vt$ (V constant) and x_2 only, then it is possible to introduce new independent variables x_1 and x_2 such that

$$X_1 = x_1 + Vt, \qquad X_2 = x_2, \qquad (2.6.2)$$
$$u_k = u_k(X_1, X_2), \qquad \sigma_{ij} = \sigma_{ij}(X_1, X_2). \qquad (2.6.3)$$

Referred to these new variables the equations (2.6.1) take the form

$$d_{ijkl} \frac{\partial^2 u_k}{\partial X_l \, \partial X_j} = 0, \tag{2.6.4}$$

where

$$d_{ijkl} = \begin{cases} c_{ijkl} - \rho V^2 \, \delta_{ik} & \text{for} \quad j = l = 1 \\ c_{ijkl} & \text{when} \quad j \neq 1 \quad \text{or} \quad l \neq 1. \end{cases} \tag{2.6.5}$$

Hence the system (1.1.1) is applicable with $N = 3$ provided the constant V is such that the ellipticity condition (1.1.3) is satisfied. All the results of the previous Sections 2.4 and 2.5 are applicable provided the constants c_{ijkl} occurring in these sections are replaced by the constants d_{ijkl} as defined by (2.6.5). Note, however, that the stress—displacement relation $\sigma_{ij} = c_{ijkl} \, \partial u_k / \partial x_l$ remains unaltered and as a consequence the c_{ijkl} are not replaced by d_{ijkl} in (2.4.12). Note, also, that not all of the symmetry properties of the c_{ijkl} carry over to the constants d_{ijkl}.

2.7 Transversely isotropic and orthotropic materials

A traversely isotropic material is characterized by five elastic constants which may be denoted by A, N, F, C and L. If the x_3-axis is such that it is normal to the transverse planes then each of the planes $x_i = 0$, $i = 1, 2, 3$ are planes of elastic symmetry and the non-zero c_{ijkl} are

$$\begin{aligned} c_{1111} &= c_{2222} = A, & c_{1122} &= N, & c_{1133} &= c_{2233} = F, \\ c_{1313} &= c_{2323} = L, & c_{1212} &= \tfrac{1}{2}(A - N), & c_{3333} &= C. \end{aligned} \tag{2.7.1}$$

The c_{ijkl} referred to any Cartesian frame of reference may be readily obtained from (2.7.1) by the transformation law for fourth order Cartesian tensors. For example, after a positive rotation of α about the x_2-axis followed by a positive rotation of θ about the x_1-axis the c_{ijkl} referred to the new rotated frame are given by

$$c'_{ijkl} = a_{ip} a_{jq} a_{kr} a_{ls} c_{pqrs}, \tag{2.7.2}$$

where

$$[a_{ij}] = \begin{bmatrix} \cos \alpha & 0 & -\sin \alpha \\ \sin \alpha \sin \theta & \cos \theta & \cos \alpha \sin \theta \\ \cos \theta \sin \alpha & -\sin \theta & \cos \alpha \cos \theta \end{bmatrix}. \tag{2.7.3}$$

When $\alpha = \theta = 0$ the roots of the sextic (2.4.11) are equal and the analysis of Sections 1.2–1.3 no longer applies. However when $\alpha = \pi/2$ and $\theta = 0$ (2.7.2) may be used to show that the non-zero c_{ijkl} of interest are given by

(omitting the primes)

$$c_{1111} = C, \quad c_{1122} = F, \quad c_{2222} = A, \quad c_{1133} = F,$$
$$c_{2233} = N, \quad c_{1313} = L, \quad c_{1212} = L, \quad c_{2323} = \tfrac{1}{2}(A - N). \tag{2.7.4}$$

The sextic (2.4.11) becomes, in this case,

$$[\tfrac{1}{2}(A - N)\tau^2 + L][AL\tau^4 - (F^2 + 2FL - AC)\tau^2 + CL] = 0. \tag{2.7.5}$$

Let

$$\tau_1^2 = -2L/(A - N)$$

so that τ_2^2 and τ_3^2 are the roots of the quartic factor in (2.7.5). Substitution in (2.4.10) now yields

$$[A_{k\alpha}] = \begin{bmatrix} 0 & \dfrac{-i(F+L)\tau_2}{C+L\tau_2^2} & \dfrac{-i(F+L)\tau_3}{C+L\tau_3^2} \\ 0 & i & i \\ 1 & 0 & 0 \end{bmatrix}. \tag{2.7.6}$$

Hence

$$[L_{i2\alpha}] = \begin{bmatrix} 0 & iL\left[\dfrac{C-F\tau_2^2}{C+L\tau_2^2}\right] & iL\left[\dfrac{C-F\tau_3^2}{C+L\tau_3^2}\right] \\ 0 & i\tau_2\left[A-\dfrac{F(F+L)}{C+L\tau_2^2}\right] & i\tau_3\left[A-\dfrac{F(F+L)}{C+L\tau_3^2}\right] \\ \tfrac{1}{2}\tau_1(A-N) & 0 & 0 \end{bmatrix}. \tag{2.7.7}$$

$$[N_{\alpha j}] = |A_{k\alpha}|^{-1} \begin{bmatrix} 0 & (F+L)\left[\dfrac{\tau_2}{C+L\tau_2^2}-\dfrac{\tau_3}{C+L\tau_3^2}\right] \\ i & \dfrac{i(F+L)\tau_3}{C+L\tau_3^2} & 0 \\ -i & \dfrac{-i(F+L)\tau_2}{C+L\tau_2^2} & 0 \end{bmatrix}. \tag{2.7.8}$$

$$[M_{\alpha j}] =$$

$$\begin{bmatrix} 0 & 0 & [\tfrac{1}{2}\tau_1(A-N)]^{-1} \\ \dfrac{i\tau_1\tau_3(A-N)}{2|L_{i2\alpha}|}\left[A-\dfrac{F(F+L)}{C+L\tau_3^2}\right] & \dfrac{-i\tau_1 L(A-N)}{2|L_{i2\alpha}|}\left[\dfrac{C-F\tau_3^2}{C+L\tau_3^2}\right] & 0 \\ \dfrac{-i\tau_1\tau_2(A-N)}{2|L_{i2\alpha}|}\left[A-\dfrac{F(F+L)}{C+L\tau_2^2}\right] & \dfrac{i\tau_1 L(A-N)}{2|L_{i2\alpha}|}\left[\dfrac{C-F\tau_2^2}{C+L\tau_2^2}\right] & 0 \end{bmatrix}.$$

$$\tag{2.7.9}$$

Similar expressions may be readily obtained for the matrices $[C_{ij}]$ and $[B_{ij}]$.

28   PHYSICAL PROBLEMS GOVERNED BY THE SYSTEM

Perhaps the simplest expressions for the matrices $[A_{k\alpha}]$, $[L_{i2\alpha}]$, etc. are obtained in the case of quasi-static deformations of a traversely isotropic material in which the x_3-axis is normal to the transverse planes. In this case the d_{ijkl} of (2.6.5) are

$$
\left.
\begin{aligned}
&d_{1111} = c_{1111} - \rho V^2 = A - \rho V^2, \quad d_{2222} = c_{2222} = A, \\
&d_{1122} = c_{1122} = N, \quad d_{1133} = c_{1133} = F, \quad d_{2233} = c_{2233} = F, \\
&d_{1313} = c_{1313} = L, \quad d_{2323} = c_{2323} = L, \quad d_{3131} = c_{3131} - \rho V^2 = L - \rho V^2, \\
&d_{1212} = c_{1212} = \tfrac{1}{2}(A - N), \quad d_{2121} = c_{2121} - \rho V^2 = \tfrac{1}{2}(A - N) - \rho V^2, \\
&d_{3333} = c_{3333} = C.
\end{aligned}
\right\}
$$

$$(2.7.10)$$

Substitution of these expressions into (2.5.40) and (2.5.41) (with the c_{ijkl} replaced by d_{ijkl} in both (2.5.40) and (2.5.41)) yields

$$
\begin{bmatrix}
A - \rho V^2 + \tfrac{1}{2}(A - N)\tau^2 & \tfrac{1}{2}(A + N)\tau & 0 \\
\tfrac{1}{2}(A + N)\tau & \tfrac{1}{2}(A - N) - \rho V^2 + A\tau^2 & 0 \\
0 & 0 & L - V^2 + L\tau^2
\end{bmatrix}
\begin{bmatrix}
A_1 \\
A_2 \\
A_3
\end{bmatrix}
$$

$$= 0, \quad (2.7.11)$$

$$
[L\tau^2 + L - \rho V^2][\tfrac{1}{2}(A - N)\tau^2 + \tfrac{1}{2}(A - N) - \rho V^2][A\tau^2 + A - \rho V^2] = 0.
$$

$$(2.7.12)$$

Let

$$
\tau_1 = i[1 - \rho V^2 / L]^{1/2}, \quad \tau_2 = i[1 - 2\rho V^2 / (A - N)]^{1/2},
$$

$$
\tau_3 = i[1 - \rho V^2 / A]^{1/2}. \quad (2.7.13)
$$

A suitable choice of $A_{k\alpha}$ is

$$
[A_{k\alpha}] =
\begin{bmatrix}
0 & (A - N - 2\rho V^2)^{1/2} & A^{1/2} \\
0 & i(A - N)^{1/2} & i(A - \rho V^2)^{1/2} \\
1 & 0 & 0
\end{bmatrix}.
\quad (2.7.14)
$$

Hence

$$[L_{i2\alpha}] =$$

$$
\begin{bmatrix}
0 & i(A - N)^{1/2}(A - N - \rho V^2) & i(A - N)(A - \rho V^2)^{1/2} \\
0 & -(A - N)(A - N - 2\rho V^2)^{1/2} & -A^{1/2}(A - N - \rho V^2) \\
iLA^{-1/2}(A - \rho V^2)^{1/2} & 0 & 0
\end{bmatrix}.
$$

$$(2.7.15)$$

The matrix $[L_{i2\alpha}]$ may be inverted to give $M_{\alpha j}$ in the form

$$M_{11} = M_{12} = M_{23} = M_{33} = 0$$

$$\left.\begin{aligned}
|L_{i2\alpha}|\, M_{13} &= -iA^{1/2}(A-N)^{1/2}(A-N-\rho V^2)^2 \\
&\quad +i(A-N)^2(A-N-2\rho V^2)^{1/2}(A-\rho V^2)^{1/2}, \\
|L_{i2\alpha}|\, M_{21} &= -iL(A-N-\rho V^2)(A-\rho V^2)^{1/2}, \\
|L_{i2\alpha}|\, M_{22} &= LA^{-1/2}(A-N)(A-\rho V^2), \\
|L_{i2\alpha}|\, M_{31} &= iLA^{-1/2}(A-N)(A-N-2\rho V^2)^{1/2}(A-\rho V^2)^{1/2}, \\
|L_{i2\alpha}|\, M_{32} &= -LA^{-1/2}(A-N)^{1/2}(A-N-\rho V^2)(A-\rho V^2)^{1/2}.
\end{aligned}\right\} \quad (2.7.16)$$

Use of (2.4.14) now yields

$$B_{13} = B_{31} = B_{23} = B_{32} = 0,$$

$$\left.\begin{aligned}
|L_{i2\alpha}|\, B_{11} &= iL\rho V^2(A-N-2\rho V^2)^{1/2}(A-\rho V^2)^{1/2}, \\
|L_{i2\alpha}|\, B_{12} &= LA^{-1/2}(A-N)^{1/2}(A-\rho V^2)^{1/2}[(A-N)^{1/2}(A-\rho V^2)^{1/2} \\
&\quad \times (A-N-2\rho V^2)^{1/2} - A^{1/2}(A-N-\rho V^2)], \\
|L_{i2\alpha}|\, B_{21} &= L(A-N)^{1/2}(A-\rho V^2)^{1/2}[A-N-\rho V^2 \\
&\quad -(A-N)^{1/2}A^{-1/2}(A-N-2\rho V^2)^{1/2}(A-\rho V^2)^{1/2}], \\
|L_{i2\alpha}|\, B_{22} &= iLA^{-1/2}\rho V^2(A-N)^{1/2}(A-\rho V^2), \\
|L_{i2\alpha}|\, B_{33} &= -i[A^{1/2}(A-N)^{1/2}(A-N-\rho V^2)^2 \\
&\quad +(A-N)^2(A-N-2\rho V^2)^{1/2}(A-\rho V^2)^{1/2}].
\end{aligned}\right\}$$

$$(2.7.17)$$

Similar expressions may be readily obtained for the elements of the matrices $[N_{k\alpha}]$ and $[C_{ij}]$.

Note that if $V = 0$ then the roots of the sextic are equal and the analysis of Sections 1.2–1.3 is not applicable.

An orthotropic material has elastic symmetry with respect to three orthogonal planes. If the Cartesian axes are so aligned that the material has elastic symmetry with respect to the three planes $x_i = 0$, $i = 1, 2, 3$, then the non-zero c_{ijkl} are

$$c_{1111}, c_{1122}, c_{1133}, c_{2222}, c_{2233}, c_{3333}, c_{2323}, c_{1313}, c_{1212}. \quad (2.7.18)$$

In this case (2.5.40) and (2.5.41) apply and hence it is possible to write down explicit expressions for the matrices $[A_{k\alpha}]$, $[L_{i2\alpha}]$, etc. For a general orientation of the Cartesian axes the c_{ijkl} will be given in terms of (2.7.18) by the tensor transformation law (2.7.2). The roots of the sextic (2.4.11) will then, in general, need to be found numerically.

3

Boundary-value problems

3.1 Introduction

The general solution (1.2.4) of the system (1.1.1) may be used to construct solutions to a number of particular boundary-value problems. Problems involving a half-plane, the cut plane and a strip bounded by two parallel lines will be considered in this chapter. In several important cases the analytical part of the solution will be obtained in closed form. However, even in such cases, the solution involves the roots of the polynomial (1.2.3) and, in general, these can only be obtained numerically.

Complex variable and integral transform techniques are used extensively in this chapter. An excellent introduction to the complex variable techniques is available in England [16] while Sneddon [34] gives an account of the required results for integral transforms.

3.2 First problem for a half-plane

Let the region R consist of the half-plane $x_2 < 0$. It is required to find a solution to the system (1.1.1) which is valid in R, tends to zero as $|z| = |x_1 + ix_2| \to \infty$ and satisfies the boundary condition

$$\phi_k(x_1, 0) = g_k(x_1), \qquad (3.2.1)$$

where $g_k(x_1)$ is known.

To construct the solution to this problem it is appropriate to employ the representations (1.2.13)–(1.2.16) for ϕ_k and ψ_{ij}. Now the $\theta_j(z)$ occurring in these equations are defined and analytic in the region under consideration, i.e. $x_2 < 0$. In the region $x_2 > 0$ they are undefined. It will prove useful to extend the definition of the $\theta_j(z)$ as follows:

$$\theta_j(z) = -\overline{\theta_j(\bar{z})}, \qquad \theta_j'(z) = -\overline{\theta_j'(\bar{z})} \quad \text{for} \quad x_2 > 0, \tag{3.2.2}$$

and hence

$$\theta_j(\bar{z}) = -\overline{\theta_j(z)}, \qquad \theta_j'(\bar{z}) = -\overline{\theta_j'(z)} \quad \text{for} \quad x_2 < 0. \tag{3.2.3}$$

Hence (1.2.13) and (1.2.14) become

$$\phi_k = \sum_\alpha A_{k\alpha} N_{\alpha j} \theta_j(z_\alpha) - \sum_\alpha \bar{A}_{k\alpha} \bar{N}_{\alpha j} \theta_j(\bar{z}_\alpha) \quad \text{for} \quad x_2 < 0, \tag{3.2.4}$$

$$\psi_{ij} = \sum_\alpha L_{ij\alpha} N_{\alpha k} \theta_k'(z_\alpha) - \sum_\alpha \bar{L}_{ij\alpha} \bar{N}_{\alpha k} \theta_k'(\bar{z}_\alpha) \quad \text{for} \quad x_2 < 0. \tag{3.2.5}$$

On the boundary $x_2 = 0$ these reduce to

$$\phi_k = \theta_k^-(x_1) - \theta_k^+(x_1), \tag{3.2.6}$$

$$\psi_{i2} = C_{ik} \theta_k'^-(x_1) - \bar{C}_{ik} \theta_k'^+(x_1), \tag{3.2.7}$$

where

$$\lim_{x_2 \to 0+} \theta_k(z) = \theta_k^+(x_1), \qquad \lim_{x_2 \to 0-} \theta_k(z) = \theta_k^-(x_1).$$

From (3.2.1) and (3.2.6) it follows that

$$\theta_k^+(x_1) - \theta_k^-(x_1) = -g_k(x_1). \tag{3.2.8}$$

Use of Cauchy's theorem now yields the analytic function $\theta_k(z)$ in the form

$$\theta_k(z) = \frac{-1}{2\pi i} \int_{-\infty}^{\infty} \frac{g_k(t)\,dt}{t - z}. \tag{3.2.9}$$

In particular, if

$$g_k(x_1) = \begin{cases} g_k \ (\text{constant}) & \text{for} \quad a < x_1 < b, \\ 0 & \text{for} \quad x_1 < a \quad \text{and} \quad x_1 > b, \end{cases} \tag{3.2.10}$$

then

$$\theta_k(z) = \frac{-g_k}{2\pi i} \log\left(\frac{z - b}{z - a}\right). \tag{3.2.11}$$

Substitution into (3.2.4) and (3.2.5) now yields expressions for ϕ_k and ψ_{ij} at all points of the half-plane $x_2 < 0$.

3.3   Second problem for a half-plane

Suppose that P_i is specified on the boundary $x_2 = 0$ of the half-plane $x_2 < 0$. Let

$$P_i(x_1, 0) = p_i(x_1), \tag{3.3.1}$$

where $p_i(x_1)$ is known. It is required to find a solution to (1.1.1) which is valid in $x_2 < 0$, satisfies the boundary condition (3.3.1) and tends to zero as $|z| \to \infty$.

To solve this problem it is advantageous to choose the representations (1.2.21)–(1.2.24) for ϕ_k and ψ_{ij}. Also, as in the previous section, it is useful to extend the definitions of the analytic functions $\chi_j(z)$, occurring in these equations, to the region $x_2 > 0$ as follows

$$\chi_j(z) = -\overline{\chi_j(\bar{z})}, \qquad \chi_j'(z) = -\overline{\chi_j'(\bar{z})} \quad \text{for} \quad x_2 > 0, \tag{3.3.2}$$

and hence

$$\chi_j(\bar{z}) = -\overline{\chi_j(z)}, \qquad \chi_j'(\bar{z}) = -\overline{\chi_j'(z)} \quad \text{for} \quad x_2 < 0. \tag{3.3.3}$$

Hence (1.2.21) and (1.2.22) become

$$\phi_k = \sum_\alpha A_{k\alpha} M_{\alpha j} \chi_j(z_\alpha) - \sum_\alpha \bar{A}_{k\alpha} \bar{M}_{\alpha j} \chi_j(\bar{z}_\alpha) \quad \text{for} \quad x_2 < 0, \tag{3.3.4}$$

$$\psi_{ij} = \sum_\alpha L_{ij\alpha} M_{\alpha k} \chi_k'(\dot{z}_\alpha) - \sum_\alpha \bar{L}_{ij\alpha} \bar{M}_{\alpha k} \chi_k'(\bar{z}_\alpha) \quad \text{for} \quad x_2 < 0. \tag{3.3.5}$$

On $x_2 = 0$ these become

$$\phi_k = B_{kj} \chi_j^-(x_1) - \bar{B}_{kj} \chi_j^+(x_1), \tag{3.3.6}$$

$$\psi_{i2} = \chi_i'^-(x_1) - \chi_i'^+(x_1). \tag{3.3.7}$$

Now the boundary of the half-plane has normal $\mathbf{n} = (0, 1)$ so that (1.2.6) and (3.3.1) yield

$$\psi_{i2}(x_1, 0) = p_i(x_1). \tag{3.3.8}$$

Hence

$$\chi_i'^+(x_1) - \chi_i'^-(x_1) = -p_i(x_1). \tag{3.3.9}$$

Use of Cauchy's theorem permits $\chi_i'(z)$ to be expressed in the form

$$\chi_i'(z) = \frac{-1}{2\pi i} \int_{-\infty}^\infty \frac{p_i(t)\,dt}{t - z}. \tag{3.3.10}$$

Substitution into (3.3.4) and (3.3.5) now yields ϕ_k and ψ_{ij} throughout the half-plane $x_2 < 0$.

Note that from (3.3.10)

$$\chi_i'(z) = \frac{1}{2\pi i z} \int_{-\infty}^\infty p_i(t)\,dt + O(z^{-2}) \tag{3.3.11}$$

so that $\chi_i(z)$ will only tend to zero as $|z|$ tends to infinity if

$$\int_{-\infty}^\infty p_i(t)\,dt = 0. \tag{3.3.12}$$

For this particular problem this is just the condition (1.1.6). However, it should be noted that the solution presented in this section is still considered to have some merit in physical applications even when (3.3.12) is not satisfied. The functions $\chi_i(z)$ and hence the ϕ_k in (3.3.4) then exhibit logarithmic behaviour for large $|z|$.

3.4   Mixed boundary-value problem for a half-plane I

Suppose that on the boundary $x_2 = 0$ of the half-plane $x_2 < 0$ the following mixed boundary conditions are required to hold:

$$P_i(x_1, 0) = p_i(x_1) \quad \text{for} \quad a < x_1 < b, \tag{3.4.1}$$

$$\phi_k(x_1, 0) = g_k(x_1) \quad \text{for} \quad x_1 < a \quad \text{and} \quad x_1 > b, \tag{3.4.2}$$

where $g_i(x_1)$ and $p_i(x_1)$ are known functions and a and b are constants. An appropriate representation for ϕ_k and ψ_{ij} for this problem is given by (3.2.4)–(3.2.7). Hence, from (3.2.6), (3.2.7), (3.4.1) and (3.4.2) it follows that

$$\bar{C}_{ik}\theta_k'^{+}(x_1) - C_{ik}\theta_k'^{-}(x_1) = -p_i(x_1) \quad \text{for} \quad a < x_1 < b, \tag{3.4.3}$$

$$\theta_k'^{+}(x_1) - \theta_k'^{-}(x_1) = -g_k'(x_1) \quad \text{for} \quad x_1 < a \quad \text{and} \quad x_1 < b. \tag{3.4.4}$$

To find the required analytical functions $\theta_k(z)$, (3.4.3) is multiplied by the constants R_i to yield

$$R_i\bar{C}_{ik}\theta_k'^{+}(x_1) - R_iC_{ik}\theta_k'^{-}(x_1) = -R_ip_i(x_1) \quad \text{for} \quad a < x_1 < b. \tag{3.4.5}$$

The R_1 are chosen such that

$$R_i\bar{C}_{ik} = S_k, \tag{3.4.6}$$

$$R_iC_{ik} = \lambda S_k. \tag{3.4.7}$$

Elimination of the S_k yields

$$(C_{ik} - \lambda\bar{C}_{ik})R_i = 0. \tag{3.4.8}$$

These equations have a non-trivial solution if

$$|C_{ik} - \lambda\bar{C}_{ik}| = 0 \tag{3.4.9}$$

which is a polynomial equation of degree N with roots which may be denoted by $\lambda_\gamma(\gamma = 1, 2, \ldots, N)$; the corresponding values of R_i and S_i obtained from (3.4.6)–(3.4.8) will be denoted by $R_{\gamma i}$ and $S_{\gamma i}$. Equation (3.4.5) may now be written

$$\{S_{\gamma k}\theta_k'^{+}(x_1)\} - \lambda_\gamma\{S_{\gamma k}\theta_k'^{-}(x_1)\} = -R_{\gamma i}p_i(x_1) \quad \text{for} \quad a < x_1 < b. \tag{3.4.10}$$

In order to find suitable $\theta_k(z)$ to satisfy (3.4.4) and (3.4.10) it is convenient to put

$$S_{\gamma k}\theta_k'(z) = X_\gamma(z)\Psi_\gamma(z) \qquad (3.4.11)$$

where $\Psi_\gamma(z)$ is an as yet undetermined function and $X_\gamma(z)$ is defined by

$$X_\gamma(z) = (z-b)^{m-1}(z-a)^{-m}, \qquad (3.4.12)$$

with

$$m = \frac{1}{2\pi i}\log \lambda_\gamma, \qquad (3.4.13)$$

where the branch of $X_\gamma(z)$ is selected so that $zX_\gamma(z) \to 1$ as $|z| \to \infty$ and the argument of λ_γ is chosen to lie between 0 and 2π. Hence (3.4.4) and (3.4.10) now become

$$\Psi_\gamma^+(x_1) - \Psi_\gamma^-(x_1) = \begin{cases} -S_{\gamma k}g_k'(x_1)[X_\gamma(x_1)]^{-1}, \\ -R_{\gamma i}p_i(x_1)[X_\gamma^+(x_1)]^{-1}, \\ -S_{\gamma k}g_k'(x_1)[X_\gamma(x_1)]^{-1}. \end{cases} \qquad (3.4.14)$$

Hence

$$\Psi_\gamma(z) = \frac{-S_{\gamma k}}{2\pi i}\int_{-\infty}^{a}\frac{g_k'(t)\,dt}{X_\gamma(t)(t-z)} - \frac{R_{\gamma i}}{2\pi i}\int_{a}^{b}\frac{p_i(t)\,dt}{X_\gamma^+(t)(t-z)}$$
$$- \frac{S_{\gamma k}}{2\pi i}\int_{b}^{\infty}\frac{g_k'(t)\,dt}{X_\gamma(t)(t-z)} + K_\gamma, \qquad (3.4.15)$$

where the K_γ are constants. Hence, from (3.4.11)

$$S_{\gamma k}\theta_k'(z) = -\frac{S_{\gamma k}X_\gamma(z)}{2\pi i}\int_{-\infty}^{a}\frac{g_k'(t)\,dt}{X_\gamma(t)(t-z)} - \frac{R_{\gamma i}X_\gamma(z)}{2\pi i}\int_{a}^{b}\frac{p_i(t)\,dt}{X_\gamma^+(t)(t-z)}$$
$$- \frac{S_{\gamma k}X_\gamma(z)}{2\pi i}\int_{b}^{\infty}\frac{g_k'(t)\,dt}{X_\gamma(t)(t-z)} + K_\gamma X_\gamma(z). \qquad (3.4.16)$$

From (3.4.16) it is apparent that, for large $|z|$

$$S_{\gamma k}\theta_k'(z) = K_\gamma z^{-1} + O(z^{-2}). \qquad (3.4.17)$$

Hence K_γ must be chosen to be zero if a solution is required for which $\theta_k(z)$ tends to zero as $|z|$ tends to infinity. However, as has already been indicated, solutions for which $\theta_k(z)$ has logarithmic behaviour at infinity are sometimes considered to give useful information in physical applications.

Finally, since the matrix $[C_{ik}]$ is non-singular (see Theorem (1.3.6)) it follows that it is possible to choose N linearly independent vectors

$R_{1i}, R_{2i}, \ldots, R_{Ni}$ corresponding to the $\lambda_1, \lambda_2, \ldots, \lambda_N$ and hence the matrix $[R_{\gamma i}]$ is non-singular. It immediately follows from (3.4.6) that the matrix $[S_{\gamma k}]$ is non-singular and hence it is possible to construct the inverse matrix $[T_{i\gamma}]$ such that

$$\sum_\gamma T_{i\gamma} S_{\gamma k} = \delta_{ik}. \tag{3.4.18}$$

Hence (3.4.16) may be readily used in conjunction with (3.4.18) to yield an explicit expression for $\theta'_k(z)$.

3.5 Mixed boundary-value problem for a half-plane II

The boundary conditions on the boundary $x_2 = 0$ of the half-plane $x_2 < 0$ are now taken in the form

$$P_i(x_1, 0) = p_i(x_1) \quad \text{for} \quad x_1 < a \quad \text{and} \quad x_1 > b, \tag{3.5.1}$$

$$\phi_k(x_1, 0) = g_k(x_1) \quad \text{for} \quad a < x_1 < b, \tag{3.5.2}$$

where, again, $p_i(x_1)$ and $g_i(x_1)$ are known functions. An appropriate representation for ϕ_k and ψ_{ij} for this problem is given by (3.3.4)–(3.3.7). Hence, from (3.3.6), (3.3.7), (3.5.1) and (3.5.2) it follows that

$$\begin{aligned} \bar{B}_{kj}\chi_j'^+(x_1) - B_{kj}\chi_j'^-(x_1) &= -g_k'(x_1) \quad \text{for} \quad a < x_1 < b, \\ \chi_i'^+(x_1) - \chi_i'^-(x_1) &= -p_i(x_1) \qquad \text{for} \quad x_1 < a \quad \text{and} \quad x_1 > b. \end{aligned} \tag{3.5.3}$$

These equations are identical in form to (3.4.3) and (3.4.4) and hence the same procedure as was used in that section may be employed to find the analytic function $\chi_i(z)$. Specifically

$$\begin{aligned} S_{\gamma k}\chi_k'(z) = &-\frac{S_{\gamma k}X_\gamma(z)}{2\pi i}\int_{-\infty}^a \frac{p_k(t)\,dt}{X_\gamma(t)(t-z)} - \frac{R_{\gamma k}X_\gamma(z)}{2\pi i}\int_a^b \frac{g_k'(t)\,dt}{X_\gamma^+(t)(t-z)} \\ &-\frac{S_{\gamma k}X_\gamma(z)}{2\pi i}\int_b^\infty \frac{p_k(t)\,dt}{X_\gamma(t)(t-z)} + K_\gamma X_\gamma(z), \end{aligned} \tag{3.5.4}$$

where

$$R_{\gamma i}\bar{B}_{ik} = S_{\gamma k}, \tag{3.5.5}$$

$$R_{\gamma i}B_{ik} = \lambda_\gamma S_{\gamma k}, \tag{3.5.6}$$

with

$$(B_{ij} - \lambda_\gamma \bar{B}_{ij})R_{\gamma i} = 0 \tag{3.5.7}$$

and the λ_γ, $\gamma = 1, 2, \ldots, N$, are the roots of the equation

$$|B_{ik} - \lambda \bar{B}_{ik}| = 0. \tag{3.5.8}$$

Also

$$X_\gamma(z) = (z-b)^{m-1}(z-a)^{-m},\qquad(3.5.9)$$

with

$$m = \frac{1}{2\pi i}\log\lambda_\gamma.\qquad(3.5.10)$$

Since the matrix $[B_{kj}]$ is non-singular (see Theorem (1.3.6)) the matrix $[R_{\gamma i}]$ may also be taken to be non-singular and hence $[S_{\gamma k}]$ is non-singular. The inverse matrix $[T_{k\gamma}]$ may therefore be defined according to (3.4.19) and an explicit expression for $\chi'_k(z)$ thereby obtained from (3.5.4).

3.6 A particular class of mixed boundary-value problems

Consider the half-plane $x_2 < 0$ with the boundary $x_2 = 0$ subject to the condition

$$P_i(x_1, 0) = 0 \quad \text{for} \quad x_1 < a \quad \text{and} \quad x_1 > b,\qquad(3.6.1)$$

where a and b are constant with $a < b$. In the interval $a < x_1 < b$ various different boundary conditions will be considered. Before proceeding to the solution of specific problems it is first observed that the appropriate representations for ϕ_k and ψ_{ij} for problems of this type are given by (3.3.4)–(3.3.7). Hence, on $x_2 = 0$

$$\phi_k = B_{kj}\chi_j^-(x_1) - \bar{B}_{kj}\chi_j^+(x_1),\qquad(3.6.2)$$

$$\psi_{i2} = \chi_i'^-(x_1) - \chi_i'^+(x_1).\qquad(3.6.3)$$

Thus, it is only necessary to require that $\chi'(z)$ be analytic in the whole plane cut along (a, b) and (3.6.1) is automatically satisfied.

Problem I

$$P_i(x_1, 0) = 0 \qquad \text{for} \quad i = 1, 2, \ldots, \gamma-1, \gamma+1, \ldots, N$$

$$(1 \leqslant \gamma \leqslant N),\quad(3.6.4)$$

$$P_\gamma(x_1, 0) = 0 \qquad \text{for} \quad x_1 < a, x_1 > b,\qquad(3.6.5)$$

$$\phi_\gamma(x_1, 0) = f(x_1) \quad \text{for} \quad a < x_1 < b,\qquad(3.6.6)$$

$$\int_a^b P_\gamma(t, 0)\,\mathrm{d}t = -P,\qquad(3.6.7)$$

where P is a constant.

In view of (3.6.2) and (3.6.4) it follows that $\chi_i(z) = 0$ for $i = 1, 2, \ldots, \gamma - 1, \gamma + 1, \ldots, N$ and $\chi_\gamma(z)$ satisfies the boundary conditions

$$\chi_\gamma'^+(x_1) - \chi_\gamma'^-(x_1) = 0 \qquad \text{for} \quad x_1 < a, x_1 > b, \tag{3.6.8}$$

$$\bar{B}_{\gamma\gamma}\chi_\gamma'^+(x_1) - B_{\gamma\gamma}\chi_\gamma'^-(x_1) = -f'(x_1) \quad \text{for} \quad a < x_1 < b. \tag{3.6.9}$$

The solution to this Hilbert problem is

$$\chi_\gamma'(z) = -\frac{X(z)}{2\pi i \bar{B}_{\gamma\gamma}} \int_a^b \frac{f'(t)\, dt}{X^+(t)(t-z)} + KX(z), \tag{3.6.10}$$

where K is a constant and

$$X(z) = (z-a)^{-m}(z-b)^{m-1}, \tag{3.6.11}$$

with

$$m = (\tfrac{1}{2}\pi i) \log [\bar{B}_{\gamma\gamma}/B_{\gamma\gamma}], \tag{3.6.12}$$

where the branch of $X(z)$ is selected so that $X(z)/z \to 1$ as $|z| \to \infty$ and the argument of $\bar{B}_{\gamma\gamma}/B_{\gamma\gamma}$ is chosen to lie between 0 and 2π. Now in Theorem 1.3.5 it was established that $B_{\gamma\gamma}$ has zero real part and hence $m = \tfrac{1}{2}$ so that

$$X(z) = (z-b)^{-1/2}(z-a)^{-1/2}. \tag{3.6.13}$$

The K occurring in (3.6.10) is directly related in the P in (3.6.7). Substitution of (3.6.10) into (3.6.3) and hence into (3.6.7) yields, upon integration

$$K = -\frac{P}{2\pi i}. \tag{3.6.14}$$

Note that the function $\chi_\gamma'(z)$ as given by (3.6.10) has square root singularities at $z = a$ and $z = b$. If the precise locations of a and b are unknown and a solution is required which does not have singularities at $z = a$ and $z = b$ then the coefficients of the singularities in (3.6.10) may be made equal to zero by suitably choosing a and b.

A second approach to the problem when a and b are unknown and a solution is required which has no singularities at $z = a$ and $z = b$ is to choose the solution with no singularities at these two points. The solution is then (3.6.10) with $X(z)$ given by

$$X(z) = (z-a)^{1/2}(z-b)^{1/2}. \tag{3.6.15}$$

The equations which determine a and b are then given by requiring that $\chi_\gamma'(z) \to 0$ as $|z| \to \infty$ and by imposing the condition (3.6.7).

Problem II

$$P_i(x_1, 0) = 0 \qquad \text{for} \quad x_1 < a, x_1 > b, \qquad (3.6.16)$$

$$P_i(x_1, 0) = \mu_i P_\gamma(x_1, 0) \quad \text{for} \quad i = 1, 2, \ldots, \gamma - 1, \gamma + 1, \ldots, N$$
$$(1 \leqslant \gamma \leqslant N) \quad \text{and} \quad a < x_1 < b, \qquad (3.6.17)$$

$$\phi_\gamma(x_1, 0) = f(x_1) \qquad \text{for} \quad a < x_1 < b, \qquad (3.6.18)$$

$$\int_a^b P_\gamma(x_1, 0) \, dx_1 = -P, \qquad (3.6.19)$$

where the μ_i and P are constants. To satisfy (3.6.17) it is necessary to put

$$\chi_i(z) = \mu_i \chi_\gamma(z) \quad \text{for} \quad i = 1, 2, \ldots, \gamma - 1, \gamma + 1, \ldots, N. \qquad (3.6.20)$$

If $\chi_\gamma(z)$ is required to be analytic in the whole plane cut along (a, b) then (3.6.16) is automatically satisfied while (3.6.18) yields

$$\bar{N}\chi_\gamma'^+(x_1) - N\chi_\gamma'^-(x_1) = -f'(x_1) \quad \text{for} \quad a < x_1 < b, \qquad (3.6.21)$$

where

$$N = \mu_1 B_{\gamma 1} + \mu_2 B_{\gamma 2} + \ldots + \mu_{\gamma-1} B_{\gamma\gamma-1} + B_{\gamma\gamma} + \mu_{\gamma+1} B_{\gamma\gamma+1} + \ldots \mu_N B_{\gamma N}. \qquad (3.6.22)$$

The appropriate solution to this Hilbert problem is

$$\chi_\gamma'(z) = -\frac{X(z)}{2\pi i \bar{N}} \int_a^b \frac{f'(t) \, dt}{X^+(t)(t-z)} + KX(z), \qquad (3.6.23)$$

where

$$X(z) = (z-a)^{-m}(z-b)^{m-1}, \qquad (3.6.24)$$

with

$$m = (2\pi i)^{-1} \log (N/\bar{N}), \qquad (3.6.25)$$

where the argument of $N/\bar{N}$ is chosen between 0 and 2π and we select the branch of $X(z)$ such that $zX(z) \to 1$ as $|z| \to \infty$. Also, from (3.6.23) and (3.6.19)

$$K = -\frac{P}{2\pi i}. \qquad (3.6.26)$$

Problem III

$$P_1(x_1, 0) = 0 \qquad \text{for} \quad x_1 < a, x_1 > b, \qquad (3.6.27)$$

$$\phi_1(x_1, 0) = g(x_1) \quad \text{for} \quad a < x_1 < b, \qquad (3.6.28)$$

$$P_j(x_1, 0) = p_j(x_1) \quad \text{for} \quad a < x_1 < b \quad \text{and} \quad j = 2, 3, \ldots, N, \qquad (3.6.29)$$

$$\int_a^b P_1(x_1, 0) \, dx_1 = T. \qquad (3.6.30)$$

Now from (3.6.3) it is clear that (3.6.27) will be satisfied if the $\chi_i'(z)$ are chosen to be analytic in the whole plane cut along (a, b) while (3.6.29) will be satisfied if

$$\chi_j'(z) = -\frac{1}{2\pi i} \int_a^b \frac{p_j(t)\,dt}{t-z} \quad \text{for} \quad j = 2, 3, \ldots, N. \tag{3.6.31}$$

Equations (3.6.2), (3.6.28) and (3.6.31) now yield

$$\bar{B}_{11}\chi_1'^+(x_1) - B_{11}\chi_1'^-(x_1) = f(x_1) \quad \text{for} \quad a < x_1 < b, \tag{3.6.32}$$

where

$$f(x_1) = -g'(x_1) + \sum_{\alpha=2}^{N} [\bar{B}_{1\alpha}\chi_\alpha'^+(x_1) - B_{1\alpha}\chi_\alpha'^-(x_1)], \tag{3.6.33}$$

where $\chi_2', \chi_3', \ldots, \chi_N'$ are obtained from (3.6.31). Now by Theorem 1.3.5 B_{11} has zero real part and hence the solution to the Hilbert problem (3.6.32) is

$$\chi_1'(z) = \frac{X(z)}{2\pi i \bar{B}_{11}} \int_a^b \frac{f(t)\,dt}{X^+(t)(t-z)} + KX(z), \tag{3.6.34}$$

where K is a constant and

$$X(z) = (z-a)^{-1/2}(z-b)^{-1/2}, \tag{3.6.35}$$

where the branch of $X(z)$ is selected so that $zX(z) \to 1$ as $|z| \to \infty$. Also, from (3.6.30) and (3.6.34)

$$K = \frac{T}{2\pi i}. \tag{3.6.36}$$

The same procedure may readily be modified to include the case when several of the $\phi_i(x_1, 0)$ are specified over $a < x_1 < b$.

Problem IV

$$P_1(x_1, 0) = 0 \qquad \text{for} \quad x_1 < a,\, x_1 > b, \tag{3.6.37}$$

$$P_2(x_1, 0) = -p(x_1) \qquad \text{for} \quad a < x_1 < b, \tag{3.6.38}$$

$$P_j(x_1, 0) = -\mu_j p(x_1) \quad \text{for} \quad a < x_1 < b \quad \text{and} \quad j = 3, 4, \ldots, N, \tag{3.6.39}$$

$$P_1(x_1, 0) = -\Lambda p(x_1) \qquad \text{for} \quad a < x_1 < c, \tag{3.6.40}$$

$$\phi_1(x_1, 0) = g(x_1) \qquad \text{for} \quad c < x_1 < d, \tag{3.6.41}$$

$$P_1(x_1, 0) = \Lambda p(x_1) \qquad \text{for} \quad d < x_1 < b, \tag{3.6.42}$$

$$\int_a^b P_1(x_1, 0)\,dx_1 = T, \tag{3.6.43}$$

where a, b, c, d, μ and Λ are constants and $p(x_1)$ and $g(x_1)$ are known functions of x_1. Condition (3.6.37) will be satisfied if the functions $\chi_i(z)$ are chosen to be analytic in the whole plane cut along (a, b) while (3.6.38) leads to

$$\chi_2'^+(x_1) - \chi_2'^-(x_1) = p(x_1) \quad \text{for} \quad a < x_1 < b. \tag{3.6.44}$$

Hence

$$\chi_2'(z) = \frac{1}{2\pi i} \int_a^b \frac{p(t)\, dt}{t - z}. \tag{3.6.45}$$

The condition (3.6.39) will be immediately satisfied if

$$\chi_j(z) = \mu_j \chi_2(z) \quad \text{for} \quad j = 3, 4, \ldots, N. \tag{3.6.46}$$

Now, from Theorem 1.3.5, B_{11} has zero real part and hence (3.6.40)–(3.6.42) lead to the equations

$$\chi_1'^+(x_1) - \chi_1'^-(x_1) = \Lambda p(x_1) \qquad \text{for} \quad a < x_1 < c, \tag{3.6.47}$$

$$\chi_1'^+(x_1) + \chi_1'^-(x_1) = \bar{B}_{11}^{-1} q(x_1) \quad \text{for} \quad c < x_1 < d, \tag{3.6.48}$$

$$\chi_1'^+(x_1) - \chi_1'^-(x_1) = -\Lambda p(x_1) \quad \text{for} \quad d < x_1 < b, \tag{3.6.49}$$

where

$$q(x_1) = -g'(x_1) - \sum_{\gamma=2}^{N} [B_{1\gamma}\chi_\gamma'^-(x_1) - \bar{B}_{1\gamma}\chi_\gamma'^+(x_1)]. \tag{3.6.50}$$

Now let

$$\chi_1'(z) = X(z)\Psi(z), \tag{3.6.51}$$

where $\Psi(z)$ is a function which is analytic along the whole plane cut along (a, b) and $X(z)$ is defined by

$$X(z) = (z - c)^{1/2}(z - d)^{1/2}, \tag{3.6.52}$$

where the branch of $X(z)$ is selected such that $z^{-1}X(z) \to 1$ as $|z| \to \infty$. Using (3.6.51) and (3.6.52), it is apparent that the conditions (3.6.47)–(3.6.49) will be satisfied if

$$\Psi^+(x_1) - \Psi^-(x_1) = \begin{cases} \Lambda p(x_1)[X(x_1)]^{-1} & \text{for} \quad a < x_1 < c, \\ q(x_1)[\bar{B}_{11}X^+(x_1)]^{-1} & \text{for} \quad c < x_1 < d, \\ -\Lambda p(x_1)[X(x_1)]^{-1} & \text{for} \quad d < x_1 < b. \end{cases}$$

Use of Cauchy's theorem now yields

$$2\pi i \Psi(z) = \Lambda \int_a^c \frac{p(t)\, dt}{X(t)(t-z)} + \bar{B}_{11}^{-1} \int_c^d \frac{q(t)\, dt}{X^+(t)(t-z)}$$

$$- \Lambda \int_d^a \frac{p(t)\, dt}{X(t)(t-z)}. \tag{3.6.53}$$

Since $\chi_1'(z)$ is required to tend to zero as $|z| \to \infty$ it follows from (3.6.51), (3.6.52) and (3.6.53) that

$$\Lambda \int_a^c \frac{p(t)\,dt}{X(t)} + \bar{B}_{11}^{-1} \int_c^d \frac{q(t)\,dt}{X^+(t)} - \Lambda \int_d^b \frac{p(t)\,dt}{X(t)} = 0. \tag{3.6.54}$$

Now

$$P_1(x_1, 0) = \psi_{12}(x_1, 0) = \chi_1'^-(x_1) - \chi_1'^+(x_1).$$

Use of Cauchy's theorem in conjunction with (3.6.37) thus yields

$$2\pi i \chi_1'(z) = - \int_a^b \frac{P_1(x_1, 0)\,dx_1}{x_1 - z}$$

$$= z^{-1} \int_a^b P_1(x_1, 0)\,dx_1 + O(z^{-2}) \quad \text{as} \quad |z| \to \infty. \tag{3.6.55}$$

Now (3.6.51)–(3.6.54) may also be used to obtain an expression for $\chi_1'(z)$ in terms of powers of $1/z$. Comparison of this expression with (3.6.55) and use of (3.6.43) yields

$$\int_a^b P_1(x_1, 0)\,dx_1 = T = -\Lambda \int_a^c \frac{(c - d + 2t)p(t)\,dt}{2X(t)} - \int_c^d \frac{(c - d + 2t)q(t)\,dt}{2\bar{B}_{11}X^+(t)}$$

$$+ \Lambda \int_d^b \frac{(c - d + 2t)p(t)\,dt}{2X(t)}. \tag{3.6.56}$$

Equations (3.6.54) and (3.6.56) serve to determine c and d once $a, b, T, p(x_1)$ and $g(x_1)$ are known. In general, c and d will have to be determined numerically but in certain special cases it is possible to obtain relatively simple expressions for both c and d.

It is apparent that a number of similar boundary-value problems could be solved by using the techniques employed in this section. The preceding four problems suffice to illustrate the possibilities and, further, will be useful in the specific applications discussed in the next chapter.

3.7 Problem for the cut plane

Consider a region consisting of the whole plane cut along the x_1-axis from a to b. Over the cut the $P_i(x_1, 0)$ are given. A solution to (1.1.1) is required which is continuous in the cut plane and tends to zero as $|z| \to \infty$. Further, P_i is required to be continuous across any line in the cut plane. It is advantageous to consider the regions $x_2 < 0$ and $x_2 > 0$ separately and the superscripts R and L, respectively, will be used to denote these two half-planes. Thus, the boundary conditions on $x_2 = 0$ may be written

$$\psi_{i2}^L(x_1, 0) = p_i(x_1) \qquad \text{for} \quad a < x_1 < b, \tag{3.7.1}$$

$$\psi_{i2}^R(x_1, 0) = q_i(x_1) \qquad \text{for} \quad a < x_1 < b, \tag{3.7.2}$$

$$\phi_k^L(x_1, 0) = \phi_k^R(x_1, 0) \quad \text{for} \quad x_1 < a \quad \text{and} \quad x_1 > b, \tag{3.7.3}$$

$$\psi_{i2}^L(x_1, 0) = \psi_{i2}^R(x_1, 0) \quad \text{for} \quad x_1 < a \quad \text{and} \quad x_1 > b, \tag{3.7.4}$$

where $p_i(x_1)$ and $q_i(x_1)$ are known functions of x_1. Equations (3.7.1) and (3.7.2) are just the boundary conditions over the cut while (3.7.3) and (3.7.4) express the necessary continuity conditions outside the cut.

In order to obtain the required solution it is convenient to use the representations (1.2.21) and (1.2.22) for ϕ_k and ψ_{ij}. Hence

$$\phi_k^L = \sum_\alpha A_{k\alpha} M_{\alpha j} \Psi_j(z_\alpha) + \sum_\alpha \bar{A}_{k\alpha} \bar{M}_{\alpha j} \bar{\Psi}_j(\bar{z}_\alpha) \qquad \text{for} \quad z_\alpha \in L, \tag{3.7.5}$$

$$\psi_{ij}^L = \sum_\alpha L_{ij\alpha} M_{\alpha k} \Psi'_k(z_\alpha) + \sum_\alpha \bar{L}_{ij\alpha} \bar{M}_{\alpha k} \bar{\Psi}'_k(\bar{z}_\alpha) \quad \text{for} \quad z_\alpha \in L, \tag{3.7.6}$$

$$\phi_k^R = \sum_\alpha A_{k\alpha} M_{\alpha j} \Omega_j(z_\alpha) + \sum_\alpha \bar{A}_{k\alpha} \bar{M}_{\alpha j} \bar{\Omega}_j(\bar{z}_\alpha) \qquad \text{for} \quad z_\alpha \in R, \tag{3.7.7}$$

$$\psi_{ij}^R = \sum_\alpha L_{ij\alpha} M_{\alpha k} \Omega'_k(z_\alpha) + \sum_\alpha \bar{L}_{ij\alpha} \bar{M}_{\alpha k} \bar{\Omega}'_k(\bar{z}_\alpha) \quad \text{for} \quad z_\alpha \in R. \tag{3.7.8}$$

Now, from (3.7.5) and (3.7.7) it is apparent that ϕ_k will be continuous across $x_2 = 0$ outside the cut if

$$B_{kj} \Psi_j^+(x_1) + \bar{B}_{kj} \bar{\Psi}_j^-(x_1) = B_{kj} \Omega_j^-(x_1) + \bar{B}_{kj} \bar{\Omega}_j^+(x_1)$$
$$\text{for} \quad x_1 < a \quad \text{and} \quad x_1 > b$$

or

$$B_{kj} \Psi_j^+(x_1) - \bar{B}_{kj} \bar{\Omega}_j^+(x_1) = B_{kj} \Omega_j^-(x_1) - \bar{B}_{kj} \bar{\Psi}_j^-(x_1)$$
$$\text{for} \quad x_1 < a \quad \text{and} \quad x_1 > b. \tag{3.7.9}$$

Thus if new functions $\chi_k(z)$, $k = 1, 2, \ldots, N$ are defined by

$$B_{kj} \Psi_j(z) - \bar{B}_{kj} \bar{\Omega}_j(z) = \chi_k(z) \quad \text{for} \quad z \in L, \tag{3.7.10}$$

$$B_{kj} \Omega_j(z) - \bar{B}_{kj} \bar{\Psi}_j(z) = \chi_k(z) \quad \text{for} \quad z \in R, \tag{3.7.11}$$

and if the functions $\chi_k(z)$ are required to be analytic in the whole plane cut along (a, b) then (3.7.9) is satisfied identically. Similarly, from (3.7.6) and (3.7.8) the ψ_{i2} (and hence the P_i) will be continuous across $x_2 = 0$ outside the cut if

$$\Psi'_i(z) - \bar{\Omega}'_i(z) = \theta_i(z) \quad \text{for} \quad z \in L, \tag{3.7.12}$$

$$\Omega'_i(z) - \bar{\Psi}'_i(z) = \theta_i(z) \quad \text{for} \quad z \in R, \tag{3.7.13}$$

where the functions $\theta_i(z)$ are analytic in the whole plane cut along (a, b). Differentiating (3.7.10) and substituting for $\bar{\Omega}_i(z)$ from (3.7.12) it follows that

$$(B_{kj} - \bar{B}_{kj}) \Psi'_j(z) = \chi'_k(z) - \bar{B}_{kj}\theta_j(z) \quad \text{for} \quad z \in L. \tag{3.7.14}$$

Similarly

$$(B_{kj} - \bar{B}_{kj}) \bar{\Psi}'_j(z) = \chi'_j(z) - B_{kj}\theta_j(z) \quad \text{for} \quad z \in R. \tag{3.7.15}$$

The matrix $[B_{kj} - \bar{B}_{kj}]$ is non-singular (see Theorem 1.3.7) and hence it is possible to form its inverse $[D_{jl}]$ and write

$$\Psi'_i(z) = D_{ik}\{\chi'_k(z) - \bar{B}_{kj}\theta_j(z)\} \quad \text{for} \quad z \in L, \tag{3.7.16}$$

$$\bar{\Psi}'_i(z) = -\bar{D}_{ik}\{\chi'_k(z) - B_{kj}\theta_j(z)\} \quad \text{for} \quad z \in R, \tag{3.7.17}$$

where

$$(B_{kj} - \bar{B}_{kj})D_{jl} = \delta_{kl}. \tag{3.7.18}$$

The conditions (3.7.1) and (3.7.2) together with (3.7.6) and (3.7.8) yield

$$D_{ik}\{\chi'^+_k(x_1) - \bar{B}_{kj}\theta^+_j(x_1)\} - \bar{D}_{ik}\{\chi'^-_k(x_1) - B_{kj}\theta^-_j(x_1)\}$$
$$= p_i(x_1) \quad \text{for} \quad a < x_1 < b, \tag{3.7.19}$$

$$-\bar{D}_{ik}\{\chi'^-_k(x_1) - B_{kj}\theta^-_j(x_1)\} + D_{ik}\{\chi'^+_k(x_1) - \bar{B}_{kj}\theta^+_j(x_1)\}$$
$$+ \theta^-_i(x_1) - \theta^+_i(x_1) = q_i(x_1) \quad \text{for} \quad a < x_1 < b. \tag{3.7.20}$$

Hence the problem reduces to one of finding functions $\theta_i(z)$ and $\chi_i(z)$ which are analytic in the whole plane cut along (a, b) and satisfy (3.7.19) and (3.7.20).

Subtraction of (3.7.20) from (3.7.19) yields

$$\theta^+_i(x_1) - \theta^-_i(x_1) = p_i(x_1) - q_i(x_1) \quad \text{for} \quad a < x_1 < b. \tag{3.7.21}$$

Hence

$$\theta_i(z) = \frac{1}{2\pi i} \int_a^b \frac{[p_i(t) - q_i(t)]\,dt}{t - z} \tag{3.7.22}$$

Both (3.7.19) and (3.7.20) now yield the same equation for $\chi'_k(z)$. Specifically,

$$D_{ik}\chi'^+_k(x_1) - \bar{D}_{ik}\chi'^-_k(x_1) = r_i(x_1) \quad \text{for} \quad a < x_1 < b, \tag{3.7.23}$$

where

$$r_i(x_1) = p_i(x_1) + D_{ik}\bar{B}_{kj}\theta^+_j(x_1) - \bar{D}_{ik}B_{kj}\theta^-_j(x_1), \tag{3.7.24}$$

with the $\theta_i(z)$ given by (3.7.21). It is apparent from (3.7.18) that $D_{jl} = -\bar{D}_{jl}$ so (3.7.23) may be written in the form

$$\{D_{ik}\chi'^+_k(x_1)\} + \{D_{ik}\chi'^-_k(x_1)\} = r_i(x_1) \quad \text{for} \quad a < x_1 < b. \tag{3.7.25}$$

This Hilbert problem has the solution

$$D_{ik}\chi'_k(z) = \frac{X(z)}{2\pi i} \int_a^b \frac{r_i(t)\,dt}{X^+(t)(t-z)},\tag{3.7.26}$$

where

$$X(z) = (z-a)^{-1/2}(z-b)^{-1/2},\tag{3.7.27}$$

where the branch of $X(z)$ is selected so that $zX(z) \to 0$ as $|z| \to \infty$. Note that the requirement that the solution tends to zero as $|z| \to \infty$ will only be satisfied if $p_i(x_1) = q_i(x_1)$ so that, from (3.7.22), the $\theta_i(z)$ are identically zero. If $p_i(x_1) \neq q_i(x_1)$ then the condition (1.1.6) is not satisfied and the solution θ_k exhibits logarithmic behaviour at infinity.

3.8 Two dissimilar half-planes with a cut along the join

As in the previous problem consider a region consisting of the whole plane cut along the x_1-axis from a to b with the P_i given over the cut. In the region L consisting of the half-space $x_2 > 0$ the constants in (1.1.1) are denoted by a^L_{ijkl}. In the region R consisting of the half-plane $x_2 < 0$ the constants in (1.1.1) are denoted a^R_{ijkl} where both a^L_{ijkl} and a^R_{ijkl} are defined for $i, k = 1, 2, \ldots, N$ and $j, l = 1, 2$. It is required to find a solution ϕ_i to (1.1.1) which satisfies the conditions outlined in Section 3.7. As in Section 3.7 it is useful to consider the regions L and R separately and hence the boundary conditions on $x_2 = 0$ are given by (3.7.1)–(3.7.4). Specifically,

$$\psi^L_{i2}(x_1, 0) = p_i(x_1) \qquad \text{for} \quad a < x_1 < b,\tag{3.8.1}$$

$$\psi^R_{i2}(x_1, 0) = q_i(x_1) \qquad \text{for} \quad a < x_1 < b,\tag{3.8.2}$$

$$\phi^L_k(x_1, 0) = \phi^R_k(x_1, 0) \quad \text{for} \quad x_1 < a \quad \text{and} \quad x_1 > b,\tag{3.8.3}$$

$$\psi^L_{i2}(x_1, 0) = \psi^L_{i2}(x_1, 0) \quad \text{for} \quad x_1 < a \quad \text{and} \quad x_1 > b,\tag{3.8.4}$$

where $p_i(x_1)$ and $q_i(x_1)$ are known functions of x. The appropriate representations for ϕ_k and ψ_{ij} are

$$\phi^L_k = \sum_\alpha A^L_{k\alpha} M^L_{\alpha j} \Psi_j(z_\alpha) + \sum_\alpha \bar{A}^L_{k\alpha} \bar{M}^L_{\alpha j} \bar{\Psi}_j(\bar{z}_\alpha) \quad \text{for} \quad z_\alpha \in L,\tag{3.8.5}$$

$$\psi^L_{ij} = \sum_\alpha L^L_{ij\alpha} M^L_{\alpha k} \Psi'_k(z_\alpha) + \sum_\alpha \bar{L}^L_{ij\alpha} \bar{M}^L_{\alpha k} \bar{\Psi}'_k(\bar{z}_\alpha) \quad \text{for} \quad z_\alpha \in L,\tag{3.8.6}$$

$$\phi^R_k = \sum_\alpha A^R_{k\alpha} M^R_{\alpha j} \Omega_j(z_\alpha) + \sum_\alpha \bar{A}^R_{k\alpha} \bar{M}^R_{\alpha j} \bar{\Omega}_j(\bar{z}_\alpha) \quad \text{for} \quad z_\alpha \in R,\tag{3.8.7}$$

$$\psi^R_{ij} = \sum_\alpha L^R_{ij\alpha} M^R_{\alpha k} \Omega'_k(z_\alpha) + \sum_\alpha \bar{L}^R_{ij\alpha} \bar{M}^R_{\alpha k} \bar{\Omega}'_k(\bar{z}_\alpha) \quad \text{for} \quad z_\alpha \in R.\tag{3.8.8}$$

The dependent variables ϕ_k will be continuous across $x_2 = 0$ outside the cut if

$$B_{kj}^L \Psi_j^+(x_1) + \bar{B}_{kj}^L \bar{\Psi}_j^-(x_1) = B_{kj}^R \Omega_j^-(x_1) + \bar{B}_{kj}^R \bar{\Omega}^+(x_1)$$

$$\text{for} \quad x_1 < a \quad \text{and} \quad x_1 > b,$$

or

$$B_{kj}^L \Psi_j^+(x_1) - \bar{B}_{kj}^R \bar{\Omega}_j^+(x_1) = B_{kj}^R \Omega_j^-(x_1) + \bar{B}_{kj}^L \bar{\Psi}^-(x_1)$$

$$\text{for} \quad x_1 < a \quad \text{and} \quad x_1 > b. \quad (3.8.9)$$

Hence if

$$B_{kj}^L \Psi_j(z) - \bar{B}_{kj}^R \bar{\Omega}_j(z) = \chi_k(z) \quad \text{for} \quad z \in L, \quad (3.8.10)$$

$$B_{kj}^R \Omega_j(z) - \bar{B}_{kj}^L \bar{\Psi}_j(z) = \chi_k(z) \quad \text{for} \quad z \in R, \quad (3.8.11)$$

where the functions $\chi_k(z)$ are analytic in the whole plane cut along (a, b) then (3.8.9) is satisfied. Similarly, P_i will be continuous across $x_2 = 0$ outside the cut if

$$\Psi_i'(z) - \bar{\Omega}_i'(z) = \theta_i(z) \quad \text{for} \quad z \in L, \quad (3.8.12)$$

$$\Omega_i'(z) - \bar{\Psi}_i'(z) = \theta_i(z) \quad \text{for} \quad z \in R \quad (3.8.13)$$

and the $\theta_i(z)$ are required to be analytic in the whole plane cut along (a, b). From (3.8.10) and (3.8.12) it follows that

$$(B_{kj}^L - \bar{B}_{kj}^R)\Psi_i'(z) = \chi_k'(z) - \bar{B}_{kj}^R \theta_j(z) \quad \text{for} \quad z \in L. \quad (3.8.14)$$

If the matrix $[B_{kj}^L - \bar{B}_{kj}^R]$ is non-singular then

$$\Psi_i'(z) = D_{ik}\{\chi_k'(z) - \bar{B}_{kj}^R \theta_j(z)\} \quad \text{for} \quad z \in L, \quad (3.8.15)$$

where

$$(B_{kj}^L - \bar{B}_{kj}^R)D_{jl} = \delta_{kl}. \quad (3.8.16)$$

Similarly

$$\Psi_i'(z) = -\bar{D}_{ik}\{\chi_k'(z) - B_{kj}^R \theta_j(z)\} \quad \text{for} \quad z \in R. \quad (3.8.17)$$

Use of (3.8.6), (3.8.8), (3.8.12)–(3.8.17) now permits the boundary conditions (3.8.1) and (3.8.2) to be recast in the form

$$D_{ik}\{\chi_k'^+(x_1) - \bar{B}_{kj}^R \theta_j^+(x_1)\} - \bar{D}_{ik}\{\chi_k'^-(x_1) - B_{kj}^R \theta_j^-(x_1)\} = p_i(x_1)$$

$$\text{for} \quad a < x_1 < b, \quad (3.8.18)$$

and

$$-\bar{D}_{ik}\{\chi_k'^-(x_1) - B_{kj}^R \theta_j^-(x_1)\} + D_{ik}\{\chi_k'^+(x_1) - \bar{B}_{kj}^R \theta_j^+(x_1)\} + \theta_i^-(x_1) - \theta_i^+(x_1)$$

$$= q_i(x_1) \quad \text{for} \quad a < x_1 < b. \quad (3.8.19)$$

Subtraction of (3.8.19) from (3.8.18) leads to

$$\theta_i^+(x_1) - \theta_i^-(x_1) = p_i(x_1) - q_i(x_1)$$

so that

$$\theta_i(z) = \frac{1}{2\pi i} \int_a^b \frac{p_i(t) - q_i(t)}{t - z} \, dt. \tag{3.8.20}$$

Equation (3.8.18) now yields

$$D_{ik}\chi_k'^+(x_1) - \bar{D}_{ik}\chi_k'^-(x_1) = r_i(x_1) \quad \text{for} \quad a < x_1 < b, \tag{3.8.21}$$

where

$$r_i(x_1) = p_i(x_1) + D_{ik}\bar{B}_{kj}^R\theta_j^+(x_1) - \bar{D}_{ik}B_{kj}^R\theta_j^-(x_1). \tag{3.8.22}$$

Equation (3.8.19) leads to the same equations (3.8.21) and (3.8.22) and hence need not be considered separately. Multiplying (3.8.21) by constants R_i which are yet to be determined it follows that

$$R_iD_{ik}\chi_k'^+(x_1) - R_i\bar{D}_{ik}\chi_k'^-(x_1) = R_i r_i(x_1) \quad \text{for} \quad a < x_1 < b. \tag{3.8.23}$$

The R_i are chosen such that

$$R_iD_{ik} = S_k, \tag{3.8.24}$$

$$R_i\bar{D}_{ik} = \lambda S_k, \tag{3.8.25}$$

where the S_k and λ are yet to be determined. Elimination of the S_k leads to

$$(\bar{D}_{ik} - \lambda D_{ik})R_i = 0. \tag{3.8.26}$$

These equations have a non-trivial solution if

$$|\bar{D}_{ik} - \lambda D_{ik}| = 0 \tag{3.8.27}$$

which is a polynomial of degree N is λ with roots which may be denoted by λ_γ, $\gamma = 1, 2, \ldots, N$; the corresponding values of R_i and S_i obtained from (3.8.26) and (3.8.24) will be denoted by $R_{\gamma i}$ and $S_{\gamma i}$. Equation (3.8.23) may now be written

$$\{S_{\gamma k}\chi_k'^+(x_1)\} - \lambda_\gamma\{S_{\gamma k}\chi_k'^-(x_1)\} = R_{\gamma i}r_i(x_1) \quad \text{for} \quad a < x_1 < b. \tag{3.8.28}$$

The appropriate solution to this Hilbert problem is

$$\{S_{\gamma k}\chi_k'(z)\} = \frac{X_\gamma(z)}{2\pi i} \int_a^b \frac{R_{\gamma i}r_i(t)\, dt}{X_\gamma^+(t)(t - z)}, \tag{3.8.29}$$

where

$$X_\gamma(z) = (z - a)^{-m}(z - b)^{m-1}, \tag{3.8.30}$$

with

$$m = \frac{1}{2\pi i} \log \lambda_\gamma, \tag{3.8.31}$$

where the branch of $X_\gamma(z)$ is selected so that $zX_\gamma(z) \rightarrow 1$ as $|z| \rightarrow \infty$ and the argument of λ_γ is chosen to lie between 0 and 2π. Provided the matrix $[S_{\gamma k}]$ is non-singular it is now possible to use (3.8.29) to obtain an explicit expression for $\chi'_k(z)$ in the form

$$\chi'_k(z) = \sum_\gamma \left\{ \frac{T_{k\gamma} X_\gamma(z)}{2\pi i} \int_a^b \frac{R_{\gamma i} r_i(t)\, dt}{X_\gamma^+(t)(t-z)} \right\}, \tag{3.8.32}$$

where

$$\sum_\alpha T_{i\gamma} S_{\gamma j} = S_{ij}. \tag{3.8.33}$$

3.9 Boundary-value problems for a strip

Consider the region R between the lines $x_2 = \pm h$ and suppose either ϕ_i or P_i is given on these lines. A solution to (1.1.1) is required which is valid in the strip and which satisfies the boundary conditions on the boundary $x_2 = \pm h$. It is advantageous to use the general solution (1.2.4) to (1.1.1) with the analytic functions $f_\alpha(z)$ given by

$$f_\alpha(z) = \frac{1}{2\pi i} \int_0^\infty \{E_\alpha(p) \exp(ipz_\alpha) + F_\alpha(p) \exp(-ipz_\alpha)\}\, dp + D_\alpha \tag{3.9.1}$$

where $E_\alpha(p)$ and $F_\alpha(p)$ are functions of p which will be determined by the boundary conditions and D_α is an arbitrary constant. Hence

$$\phi_k = \frac{1}{\pi} \mathcal{R} \sum_\alpha A_{k\alpha} \left\{ \int_0^\infty [E_\alpha(p) \exp(ipz_\alpha) + F_\alpha(p) \exp(-ipz_\alpha)]\, dp + D_\alpha \right\},$$

$$\tag{3.9.2}$$

$$\psi_{ij} = \frac{1}{\pi} \mathcal{R} \sum_\alpha L_{ij\alpha} \left\{ \int_0^\infty [E_\alpha(p) \exp(ipz_\alpha) - F_\alpha(p) \exp(-ipz_\alpha)]ip\, dp \right\},$$

$$\tag{3.9.3}$$

On $x_2 = \pm h$ these equations become

$$\phi_k(x_1, h) = \frac{1}{\pi} \mathscr{R}\left\{\int_0^\infty \sum_\alpha [A_{k\alpha}E_\alpha(p)\exp(ip\tau_\alpha h)\right.$$
$$\left. + \bar{A}_{k\alpha}\bar{F}_\alpha(p)\exp(ip\bar{\tau}_\alpha h)]\exp(ipx_1)\,dp + \sum_\alpha A_{k\alpha}D_\alpha\right\},$$

(3.9.4)

$$\psi_{ij}(x_1, h) = \frac{1}{\pi}\mathscr{R}\int_0^\infty \sum_\alpha [L_{ij\alpha}E_\alpha(p)\exp(ip\tau_\alpha h)$$
$$+ \bar{L}_{ij\alpha}\bar{F}_\alpha(p)\exp(ip\bar{\tau}_\alpha h)]ip\exp(ipx_1)\,dp,$$

(3.9.5)

$$\phi_k(x_1, -h) = \frac{1}{\pi}\mathscr{R}\left\{\int_0^\infty \sum_\alpha [A_{k\alpha}E_\alpha(p)\exp(-ip\tau_\alpha h)\right.$$
$$\left. + \bar{A}_{k\alpha}\bar{F}_\alpha(p)\exp(-ip\bar{\tau}_\alpha h)]\exp(ipx_1)\,dp + \sum_\alpha A_{k\alpha}D_\alpha\right\},$$

(3.9.6)

$$\psi_{ij}(x_1, -h) = \frac{1}{\pi}\mathscr{R}\int_0^\infty \sum_\alpha [L_{ij\alpha}E_\alpha(p)\exp(-ip\tau_\alpha h)$$
$$+ \bar{L}_{ij\alpha}\bar{F}_\alpha(p)\exp(-ip\bar{\tau}_\alpha h)]ip\exp(ipx_1)\,dp.$$

(3.9.7)

The boundary conditions on $x_2 = \pm h$ will only involve ψ_{i2} and hence it is convenient to write

$$\frac{1}{\pi}\mathscr{R}\left\{\int_0^\infty \sum_\alpha [U_{k\alpha}E_\alpha + \bar{V}_{k\alpha}\bar{F}_\alpha]\exp(ipx_1)\,dp + \sum_\alpha A_{k\alpha}D_\alpha\right\} = \phi_k(x_1, h),$$

(3.9.8)

$$\frac{1}{\pi}\mathscr{R}\int_0^\infty \sum_\alpha [S_{i\alpha}E_\alpha + \bar{R}_{i\alpha}\bar{F}_\alpha]ip\exp(ipx_1)\,dp = \psi_{i2}(x_1, h),$$

(3.9.9)

$$\frac{1}{\pi}\mathscr{R}\left\{\int_0^\infty \sum_\alpha [V_{k\alpha}E_\alpha + \bar{U}_{k\alpha}\bar{F}_\alpha]\exp(ipx_1)\,dp + \sum_\alpha A_{k\alpha}D_\alpha\right\} = \phi_k(x_1, -h),$$

(3.9.10)

$$\frac{1}{\pi}\mathscr{R}\int_0^\infty \sum_\alpha [R_{i\alpha}E_\alpha + \bar{S}_{i\alpha}\bar{F}_\alpha]ip\exp(ipx_1)\,dp = \psi_{i2}(x_1, -h),$$

(3.9.11)

where

$$R_{i\alpha} = L_{i2\alpha}\exp(-ip\tau_\alpha h),$$

(3.9.12)

$$S_{i\alpha} = L_{i2\alpha}\exp(ip\tau_\alpha h),$$

(3.9.13)

$$U_{k\alpha} = A_{k\alpha}\exp(ip\tau_\alpha h),$$

(3.9.14)

$$V_{k\alpha} = A_{k\alpha}\exp(-ip\tau_\alpha h).$$

(3.9.15)

If either ϕ_k or ψ_{i2} are given on the boundaries $x_2 = \pm h$ then (3.9.8)–(3.9.11) may be used to determine $E_\alpha(p)$ and $F_\alpha(p)$ by employing the inversion theorem for Fourier transforms (see Sneddon [34]). For example if the ψ_{i2} are given on $x_2 = \pm h$ then (3.9.9) and (3.9.11) yield

$$\sum_\alpha [S_{i\alpha} E_\alpha + \bar{R}_{i\alpha} \bar{F}_\alpha] = \frac{i}{p} \int_{-\infty}^\infty \psi_{i2}(\xi, h) \exp(-ip\xi) \, d\xi, \qquad (3.9.16)$$

$$\sum_\alpha [R_{i\alpha} E_\alpha + \bar{S}_{i\alpha} \bar{F}_\alpha] = \frac{i}{p} \int_{-\infty}^\infty \psi_{i2}(\xi, h) \exp(-ip\xi) \, d\xi. \qquad (3.9.17)$$

This system of equations may be readily solved for the $E_\alpha(p)$ and $F_\alpha(p)$ which may then be substituted back into (3.9.2) and (3.9.3) to yield the required ϕ_k and ψ_{ij}.

It is apparent that if (3.9.16) and (3.9.17) are used to determine the E_α and F_α then the integral in (3.9.1) will be improper. This difficulty may be circumvented by an appropriate choice of the constant D_α in (3.9.1). Specifically, the constant may be chosen so that (3.9.1) may be written in the form

$$f_\alpha(z) = \frac{1}{2\pi i} \int_0^\infty \{E_\alpha(p)[\exp(ipz_\alpha) - 1] + F_\alpha(p)[\exp(-ipz_\alpha) - 1\} \, dp. \qquad (3.9.18)$$

This form for $f_\alpha(z)$ leads to a proper integral when $E_\alpha(p)$ and $F_\alpha(p)$ are obtained from (3.9.16) and (3.9.17).

3.10   Problem for the cut strip

Consider the strip $|x_2| < h$ with a cut along the x_1-axis from $x_1 = -a$ to $x_1 = a$. On the boundary lines $x_2 = \pm h$ either ϕ_k or P_i is given while over the cut P_i is specified. Further, the P_i on one side of the cut is required to be equal in magnitude and opposite in sign to the P_i on the other side. To find a solution to (1.1.1) which satisfies these boundary conditions it is convenient to write ϕ_k and ψ_{ij} as the sum of two separate solutions. Hence

$$\phi_k = \phi_k^{(1)} + \phi_k^{(2)}, \qquad (3.10.1)$$

$$\psi_{ij} = \psi_{ij}^{(1)} + \psi_{ij}^{(2)}. \qquad (3.10.2)$$

For $\phi_k^{(1)}$ and $\psi_{ij}^{(1)}$ the expressions (3.9.2) and (3.9.3) are used so that

$$\phi_k^{(1)} = \frac{1}{\pi} \mathcal{R} \sum_\alpha A_{k\alpha} \left\{ \int_0^\infty [E_\alpha(p) \exp(ipz_\alpha) + F_\alpha(p) \exp(-ipz_\alpha)] \, dp \right\}, \qquad (3.10.3)$$

$$\psi_{ij}^{(1)} = \frac{1}{\pi} \mathcal{R} \sum_\alpha L_{ij\alpha} \left\{ \int_0^\infty [E_\alpha(p) \exp(ipz_\alpha) - F_\alpha(p) \exp(-ipz_\alpha)] ip \, dp \right\}. \qquad (3.10.4)$$

For $\phi_k^{(2)}$ and $\psi_{ij}^{(2)}$ the regions $0 < x_2 < h$ and $-h < x_2 < 0$ are considered separately. For $0 < x_2 < h$ the function $f_\alpha(z)$ in (1.2.4) is required to adopt the form

$$f_\alpha(z) = \frac{1}{2\pi} \int_0^\infty G_\alpha^+(p) \exp(ipz_\alpha) \, dp, \qquad (3.10.5)$$

so that

$$\phi_k^{(2)} = \frac{1}{\pi} \mathcal{R} \sum_\alpha A_{k\alpha} \left\{ \int_0^\infty G_\alpha^+(p) \exp(ipz_\alpha) \, dp \right\} \qquad \text{for} \quad 0 < x_2 < h,$$
$$(3.10.6)$$

$$\psi_{ij}^{(2)} = \frac{1}{\pi} \mathcal{R} \sum_\alpha L_{ij\alpha} \left\{ \int_0^\infty G_\alpha^+(p) \exp(ipz_\alpha) ip \, dp \right\} \qquad \text{for} \quad 0 < x_2 < h,$$
$$(3.10.7)$$

where $G_\alpha^+(p)$ will be determined from the boundary conditions. The corresponding expressions in $-h < x_2 < 0$ are

$$f_\alpha(z) = \frac{1}{2\pi} \int_0^\infty G_\alpha^-(p) \exp(-ipz_\alpha) \, dp, \qquad (3.10.8)$$

$$\phi_k^{(2)} = \frac{1}{\pi} \mathcal{R} \sum_\alpha A_{k\alpha} \left\{ \int_0^\infty G_\alpha^-(p) \exp(-ipz_\alpha) \, dp \right\} \qquad \text{for} \quad -h < x_2 < 0,$$
$$(3.10.9)$$

$$\psi_{ij}^{(2)} = \frac{1}{\pi} \mathcal{R} \sum_\alpha L_{ij\alpha} \left\{ \int_0^\infty G_\alpha^-(p) \exp(-ipz_\alpha) ip \, dp \right\} \qquad \text{for} \quad -h < x_2 < 0.$$
$$(3.10.10)$$

The ψ_{i2} must be continuous across the plane $x_2 = 0$ and from (3.10.4), (3.10.7) and (3.10.10) it is apparent that this will be the case if

$$\sum_\alpha L_{i2\alpha} G_\alpha^-(p) = \sum \bar{L}_{i2\alpha} \bar{G}_\alpha^-(p). \qquad (3.10.11)$$

If these expressions are denoted by $\chi_i(p)$ then it follows that

$$G_\alpha^+(p) = M_\alpha \chi_i(p), \qquad (3.10.12)$$
$$G_\alpha^-(p) = M_\alpha \bar{\chi}_i(p). \qquad (3.10.13)$$

Use of (3.10.1)–(3.10.13) now yields the following expressions for ϕ_k and ψ_{ij}:

$$\phi_k = \frac{1}{\pi} \mathcal{R} \sum_\alpha A_{k\alpha} \int_0^\infty \{ [E_\alpha(p) + M_{\alpha i} \chi_i(p)] \exp(ipz_\alpha)$$
$$+ F_\alpha(p) \exp(-ipz_\alpha) \} \, dp \quad \text{for} \quad 0 < x_2 < h, \qquad (3.10.14)$$

$$\psi_{ij} = \frac{1}{\pi} \mathscr{R} \sum_{\alpha} L_{ij\alpha} \int_0^{\infty} \{ [E_{\alpha}(p) + M_{\alpha k} \chi_k(p)] \exp(ipz_{\alpha})$$

$$- F_{\alpha}(p) \exp(-ipz_{\alpha}) \} ip \, dp \quad \text{for} \quad 0 < x_2 < h, \tag{3.10.15}$$

$$\phi_k = \frac{1}{\pi} \mathscr{R} \sum_{\alpha} A_{k\alpha} \int_0^{\infty} \{ E_{\alpha}(p) \exp(ipz_{\alpha}) + [F_{\alpha}(p) + M_{\alpha i} \bar{\chi}_i(p)] $$

$$\times \exp(-ipz_{\alpha}) \} \, dp \quad \text{for} \quad -h < x_2 < 0, \tag{3.10.16}$$

$$\psi_{ij} = \frac{1}{\pi} \mathscr{R} \sum_{\alpha} L_{ij\alpha} \int_0^{\infty} \{ E_{\alpha}(p) \exp(ipz_{\alpha}) - [F_{\alpha}(p) + M_{i\alpha} \bar{\chi}_i(p)] $$

$$\times \exp(-ipz_{\alpha}) \} ip \, dp \quad \text{for} \quad -h < x_2 < 0. \tag{3.10.17}$$

The unknowns $E_{\alpha}(p)$, $F_{\alpha}(p)$ and χ_k in these equations will be determined from the following conditions:
(1) the boundary conditions on $x_2 = \pm h$,
(2) continuity of ϕ_k outside the cut,
(3) the prescribed P_i (and hence ψ_{i2}) over the cut.
On $x_2 = \pm h$ these equations yield

$$\frac{1}{\pi} \mathscr{R} \int_0^{\infty} \sum_{\alpha} [U_{k\alpha}(E_{\alpha} + M_{\alpha i} \chi_i) + \bar{V}_{k\alpha} \bar{F}_{\alpha}] \exp(ipx_1) \, dp = \phi_k(x_1, h),$$

$$\tag{3.10.18}$$

$$\frac{1}{\pi} \mathscr{R} \int_0^{\infty} \sum_{\alpha} [S_{i\alpha}(E_{\alpha} + M_{\alpha k} \chi_k) + \bar{R}_{i\alpha} \bar{F}_{\alpha}] ip \exp(ipx_1) \, dp = \psi_{i2}(x, h),$$

$$\tag{3.10.19}$$

$$\frac{1}{\pi} \mathscr{R} \int_0^{\infty} \sum_{\alpha} [V_{k\alpha} E_{\alpha} + \bar{U}_{k\alpha}(\bar{F}_{\alpha} + \bar{M}_{\alpha i} \chi_i)] \exp(ipx_1) \, dp = \phi_k(x_1, -h),$$

$$\tag{3.10.20}$$

$$\frac{1}{\pi} \mathscr{R} \int_0^{\infty} \sum_{\alpha} [R_{i\alpha} E_{\alpha} + \bar{S}_{i\alpha}(\bar{F}_{\alpha} + \bar{M}_{\alpha k} \chi_k)] ip \exp(ipx_1) \, dp = \psi_{i2}(x_1, -h),$$

$$\tag{3.10.21}$$

where $U_{k\alpha}$, $V_{k\alpha}$, $S_{i\alpha}$ and $R_{i\alpha}$ are defined by (3.9.12)–(3.9.15). If either ϕ_k or ψ_{i2} are given on the boundaries $x_2 = \pm h$ then (3.10.18)–(3.10.21) may be used together with the inversion theorem for Fourier transforms to obtain expressions relating E_{α} and F_{α} to the unknown function χ_i. Specifically these expressions are

$$E_{\alpha} = Q_{\alpha} \chi_i + H_{\alpha}, \tag{3.10.22}$$

$$F_{\alpha} = Q_{\alpha} \bar{\chi}_i + I_{\alpha}, \tag{3.10.23}$$

where the $Q_{\alpha i}$, H_{α} and I_{α} will depend on the $U_{k\alpha}$, $V_{k\alpha}$, $S_{i\alpha}$ and $R_{i\alpha}$ while H_{α} and I_{α} will also depend on the boundary conditions on $x_2 = \pm h$. The precise forms for $Q_{\alpha i}$, H_α and I_α are readily determined once the conditions on $x_2 = \pm h$ are known.

From (3.10.14) and (3.10.16) it follows that the difference in ϕ_k across $x_2 = 0$ is

$$\Delta\phi_k = \frac{1}{\pi} \mathfrak{R}(B_{kj} - \bar{B}_{kj}) \int_0^\infty \chi_j(p) \exp(ipx_1) \, \mathrm{d}p. \tag{3.10.24}$$

The difference $\Delta\phi_k$ outside the cut must be zero. Also, the ψ_{i2} are prescribed over the cut. These conditions yield the equations

$$\mathfrak{R}(B_{kj} - \bar{B}_{kj}) \int_0^\infty \chi_j(p) \exp(ipx_1) \, \mathrm{d}p = 0 \quad \text{for} \quad |x_1| > a, \tag{3.10.25}$$

$$\mathfrak{R} \int_0^\infty [\chi_j(p) + \sum_\alpha \{L_{j2\alpha}E_\alpha(p) + \bar{L}_{j2\alpha}\bar{F}_\alpha(p)\}]ip \exp(ipx_1) \, \mathrm{d}p = \pi p_j(x_1)$$

$$\text{for} \quad |x_1| < a, \tag{3.10.26}$$

where the $\psi_j(x_1, 0) = p_j(x_1)$ are given functions of x_1. Now (3.10.25) will be satisfied if $\chi_j(p)$ is required to adopt the form

$$\chi_j(p) = \chi_j'(p) + \chi_j''(p)$$

$$= \int_0^a s_j(t)J_1(pt) \, \mathrm{d}t + i \int_0^a r_j(t)J_0(pt) \, \mathrm{d}t, \tag{3.10.27}$$

where the $r_j(t)$ and $s_j(t)$, $j = 1, 2, \ldots, N$ are real functions to be determined and J_0 and J_1 are Bessel functions of orders zero and one respectively. Use of (3.10.22) and (3.10.23) in (3.10.26) leads to the equation

$$\mathfrak{R} \int_0^\infty [\chi_j(p) + T_{jk}(p)\chi_k(p)]ip \exp(ipx_1) \, \mathrm{d}p = P_j(x_1) \quad \text{for} \quad |x_1| < a, \tag{3.10.28}$$

where

$$P_j(x_1) = \pi p_j(x_1) - \mathfrak{R} \int_0^\infty \sum_\alpha [L_{j2\alpha}H_\alpha(p) + \bar{L}_{j2\alpha}\bar{I}_\alpha(p)]ip \exp(ipx_1) \, \mathrm{d}p \tag{3.10.29}$$

and

$$T_{jk}(p) = 2\mathfrak{R} \sum_\alpha L_{j2\alpha}Q_{\alpha k}(p). \tag{3.10.30}$$

Use of (3.10.27) in (3.10.28) yields

$$\int_0^\infty p \cos(px_1)\, dp \int_0^a r_j(t) J_0(pt)\, dt + \int_0^\infty T_{jk}(p) \cos(px_1)\, dp \int_0^a r_k(t) J_0(pt)\, dt$$

$$= -\tfrac{1}{2}[P_j(x_1) + P_j(-x_1)] \quad \text{for} \quad 0 < x_1 < a, \quad (3.10.31)$$

$$\int_0^\infty p \sin(px_1)\, dp \int_0^a s_j(t) J_1(pt)\, dt + \int_0^\infty T_{jk}(p) p \sin(px_1)\, dp \int_0^a s_k(t) J_1(pt)\, dt$$

$$= -\tfrac{1}{2}[P_j(x_1) - P_j(-x_1)] \quad \text{for} \quad 0 < x_1 < a. \quad (3.10.32)$$

The order of integration in the first term on the left-hand side of (3.10.31) and (3.10.32) may be interchanged and standard results for Bessel functions used to obtain

$$\frac{d}{dx_1} \int_0^{x_1} \frac{r_j(t)\, dt}{(x_1^2 - t^2)^{1/2}} + \int_0^\infty T_{jk}(p) p \cos(px_1)\, dp \int_0^a r_j(t) J_0(pt)\, dt$$

$$= -\tfrac{1}{2}[P_j(x_1) + P_j(-x_1)] \quad \text{for} \quad 0 < x_1 < a, \quad (3.10.33)$$

$$\frac{1}{x_1} \frac{d}{dx_1} \int_0^{x_1} \frac{t s_j(t)\, dt}{(x_1^2 - t^2)^{1/2}} + \int_0^\infty T_{jk}(p) p \sin(px_1)\, dp \int_0^a s_j(t) J_1(pt)\, dt$$

$$= -\tfrac{1}{2}[P_j(x_1) - P_j(-x_1)] \quad \text{for} \quad 0 < x_1 < a. \quad (3.10.34)$$

These Abel type integral equations may be inverted to give

$$r_j(t) + \frac{2}{\pi} t \int_0^t \frac{du}{(t^2 - u^2)^{1/2}} \int_0^\infty T_{jk}(p) p \cos(pu)\, dp \int_0^a r_j(q) J_0(pq)\, dq$$

$$= -\frac{t}{\pi} \int_{-t}^t \frac{P_j(u)\, du}{(t^2 - u^2)^{1/2}} \quad \text{for} \quad 0 < t < a, \quad (3.10.35)$$

$$s_j(t) + \frac{2}{\pi} \int_0^t \frac{u\, du}{(t^2 - u^2)^{1/2}} \int_0^\infty T_{jk}(p) p \sin(pu)\, dp \int_0^a s_j(q) J_1(pq)\, dq$$

$$= -\frac{1}{\pi} \int_{-t}^t \frac{u P_j(u)\, du}{(t^2 - u^2)^{1/2}} \quad \text{for} \quad 0 < t < a. \quad (3.10.36)$$

Interchanging the order of integration and using the results

$$\frac{2}{\pi} \int_0^t \frac{\cos(pu)\, du}{(t^2 - u^2)^{1/2}} = J_0(pt), \quad (3.10.37)$$

$$\frac{2}{\pi} \int_0^t \frac{u \sin(pu)\, du}{(t^2 - u^2)^{1/2}} = t J_1(pt), \quad (3.10.38)$$

it follows that

$$r_j(t) + t \int_0^a K_{jk}^{(0)}(u, t) r_k(u) \, du = -\frac{t}{\pi} \int_{-t}^t \frac{P_j(u) \, du}{(t^2 - u^2)^{1/2}} \quad \text{for} \quad 0 < t < a,$$

(3.10.39)

$$s_j(t) + t \int_0^a K_{jk}^{(1)}(u, t) s_k(u) \, du = -\frac{1}{\pi} \int_{-t}^t \frac{u P_j(u) \, du}{(t^2 - u^2)^{1/2}} \quad \text{for} \quad 0 < t < a,$$

(3.10.40)

where

$$K_{jk}^{(N)}(u, t) = \int_0^\infty T_{jk}(p) J_N(pu) J_N(pt) p \, dp.$$

(3.10.41)

Equation (3.10.39) constitutes N simultaneous Fredholm equations for the $r_j(t)$ while (3.10.40) constitutes N similar equations for the $s_j(t)$.

3.11 Problem of a cut strip between two half-planes

Consider a region made up of the strip $-h < x_2 < h$ between the two half-planes $x_2 < -h$ and $x_2 > h$. The strip is cut along the x_1-axis from $-a$ to a. Across $x_2 = \pm h$ both ϕ_i and P_i are required to be continuous while over the two sides of the cut equal and opposite P_i are specified. In each of three regions $x_2 > h$, $x_2 < -h$ (henceforth denoted by L and R respectively) and $-h < x_2 < h$ (1.1.1) is the governing equation but the constants occurring in that equation may be different for each region. Specifically, the constants in L will be denoted by a_{ijkl}^L, in R by a_{ijkl}^R and in $-h < x_2 < h$ by a_{ijkl} where all of these constants are defined for $i, k = 1, 2, \ldots, N$ and $j, l = 1, 2$.

In order to find a suitable solution to (1.1.1) which satisfies the continuity conditions on $x_2 = \pm h$ and the boundary conditions on the cut it is advantageous to use the representations (1.2.4) and (1.2.7) for ϕ_i and ψ_{ij}, respectively. In L an appropriate form for the unknown function $f_\alpha(z)$ in these representations is

$$f_\alpha(z) = \frac{1}{2\pi} \int_0^\infty D_\alpha^L(p) \exp(ipz) \, dp,$$

(3.11.1)

where the functions D_α^L will be determined by the continuity conditions on $x_2 = h$. Hence

$$\phi_k^L = \frac{1}{\pi} \Re \sum_\alpha A_{k\alpha}^L \int_0^\infty D_\alpha^L(p) \exp(ipz_\alpha^L) \, dp,$$

(3.11.2)

$$\psi_{ij}^L = \frac{1}{\pi} \Re \sum_\alpha L_{ij\alpha} \int_0^\infty D_\alpha^L(p) \exp(ipz_\alpha^L) ip \, dp.$$

(3.11.3)

Similarly, in $x_2 < -h$ the expressions for ϕ_k and ψ_{ij} are

$$\phi_k^R = \frac{1}{\pi} \Re \sum_\alpha A_{k\alpha}^R \int_0^\infty D_\alpha^R(p) \exp{(ipz_\alpha^R)}\,dp, \tag{3.11.4}$$

$$\psi_{ij}^R = \frac{1}{\pi} \Re \sum_\alpha L_{ij\alpha}^R \int_0^\infty D_\alpha^R(p) \exp{(ipz_\alpha^R)} ip\,dp, \tag{3.11.5}$$

where the $D_\alpha^R(p)$ will be determined by the continuity conditions on $x_2 = -h$.

For $-h < x_2 < h$ the regions $-h < x_2 < 0$ and $0 < x_2 < h$ are considered separately. From the analysis of the previous section it is apparent that the following representations for ϕ_k and ψ_{ij} will be appropriate in these two regions.

In $0 < x_2 < h$:

$$\phi_k = \frac{1}{\pi} \Re \sum_\alpha A_{k\alpha} \int_0^\infty \{[E_\alpha(p) + M_{\alpha i}\chi_i(p)] \exp{(ipz_\alpha)}$$

$$+ F_\alpha(p) \exp{(-ipz_\alpha)}\}\,dp, \tag{3.11.6}$$

$$\psi_{ij} = \frac{1}{\pi} \Re \sum_\alpha L_{ij\alpha} \int_0^\infty \{[E_\alpha(p) + M_{\alpha i}\chi_i(p)] \exp{(ipz_\alpha)}$$

$$- F_\alpha(p) \exp{(-ipz_\alpha)}\} ip\,dp. \tag{3.11.7}$$

In $-h < x_2 < 0$:

$$\phi_k = \frac{1}{\pi} \Re \sum_\alpha A_{k\alpha} \int_0^\infty \{E_\alpha(p) \exp{(ipz_\alpha)}$$

$$+ [M_{\alpha i}\bar{\chi}_i(p) + F_\alpha(p)] \exp{(-ipz_\alpha)}\}\,dp, \tag{3.11.8}$$

$$\psi_{ij} = \frac{1}{\pi} \Re \sum_\alpha L_{ij\alpha} \int_0^\infty \{E_\alpha(p) \exp{(ipz_\alpha)}$$

$$- [M_{\alpha i}\bar{\chi}_i(p) + F_\alpha(p)] \exp{(-ipz_\alpha)}\} ip\,dp. \tag{3.11.9}$$

In (3.11.6)–(3.11.9) the $E_\alpha(p)$ and $F_\alpha(p)$ will be determined by the continuity conditions on $x_2 = \pm h$ while the $\chi_i(p)$ will be determined by the continuity and boundary conditions on $x_2 = 0$.

It is apparent from (3.11.7) and (3.11.9) that the stress is continuous across $x_2 = 0$. The difference in the displacement across the plane $x_2 = 0$ is (from (3.11.6) and (3.11.8))

$$\Delta\phi_k = \frac{1}{\pi} \Re(B_{kj} - \bar{B}_{kj}) \int_0^\infty \chi_i(p) \exp{(ipx_1)}\,dp. \tag{3.11.10}$$

Now Δu_k must be zero outside the crack and this condition together with the boundary conditions on the cut yields

$$\mathcal{R}(B_{kj} - \bar{B}_{kj}) \int_0^\infty \chi_j(p) \exp(ipx_1) \, dp = 0 \quad \text{for} \quad |x_1| > a, \tag{3.11.11}$$

$$\frac{1}{\pi} \mathcal{R} \int_0^\infty [\chi_j(p) + \sum_\alpha \{L_{j2\alpha} E_\alpha(p) + \bar{L}_{j2\alpha} F_\alpha(p)\}] ip \exp(ipx_1) \, dp = p_j(x_1)$$

$$\text{for} \quad |x_1| < a, \tag{3.11.12}$$

where the $p_j(x_1)$ are the given values of $\psi_{i2}(x_1, 0)$ over the cut. Both ϕ_k and P_i (and hence ψ_{i2}) must be continuous across $x_2 = \pm h$. From (3.11.2)–(3.11.9) it follows that this requirement will be satisfied if

$$\sum_\alpha [L_{i2\alpha}\{E_\alpha(p) + M_{\alpha j}\chi_j(p)\} \exp(ip\tau_\alpha h) + \bar{L}_{i2\alpha}\bar{F}_\alpha(p) \exp(ip\bar{\tau}_\alpha h)]$$

$$= \sum_\alpha L_{i2\alpha}^L D_\alpha^L(p) \exp(ip\tau_\alpha^L h), \tag{3.11.13}$$

$$\sum_\alpha [L_{i2\alpha} E_\alpha(p) \exp(-ip\tau_\alpha h) + \bar{L}_{i2\alpha}\{\bar{F}_\alpha(p) + \bar{M}_{\alpha j}\chi_j(p)\} \exp(-ip\bar{\tau}_\alpha h)]$$

$$= \sum_\alpha \bar{L}_{i2\alpha}^R \bar{D}_\alpha^R(p) \exp(ip\tau_\alpha^R h), \tag{3.11.14}$$

$$\sum_\alpha [A_{k\alpha}\{E_\alpha(p) + M_{\alpha j}\chi_j(p)\} \exp(ip\tau_\alpha h) + \bar{A}_{k\alpha}\bar{F}_\alpha(p) \exp(ip\bar{\tau}_\alpha h)]$$

$$= \sum_\alpha A_{k\alpha}^L D_\alpha^L(p) \exp(ip\tau_\alpha^L h), \tag{3.11.15}$$

$$\sum_\alpha [A_{k\alpha} E_\alpha(p) \exp(-ip\tau_\alpha h) + \bar{A}_{k\alpha}\{\bar{F}_\alpha(p) + \bar{M}_{\alpha j}\chi_j(p)\} \exp(ip\bar{\tau}_\alpha h)]$$

$$= \sum_\alpha A_{k\alpha}^k D_\alpha^R(p) \exp(ip\tau_\alpha^R h). \tag{3.11.16}$$

It is now convenient to put

$$D_\alpha^L(p) = M_{\alpha j}^L \exp(-ip\tau_\alpha^L h)\theta_j(p), \tag{3.11.17}$$

$$D_\alpha^R(p) = M_{\alpha j}^R \exp(ip\tau_\alpha^R h)\omega_j(p). \tag{3.11.18}$$

Use of these equations in (3.11.13)–(3.11.16) gives rise to the equations

$$\mathbf{SE} + \bar{\mathbf{R}}\bar{\mathbf{F}} = -\mathbf{SMX} + \mathbf{\Theta}, \tag{3.11.19}$$

$$\mathbf{RE} + \bar{\mathbf{S}}\bar{\mathbf{F}} = -\bar{\mathbf{S}}\bar{\mathbf{M}}\mathbf{X} + \mathbf{\Omega}, \tag{3.11.20}$$

$$\mathbf{UE} + \bar{\mathbf{V}}\bar{\mathbf{F}} = -\mathbf{UMX} + \mathbf{B}^L\mathbf{\Theta}, \tag{3.11.21}$$

$$\mathbf{VE} + \bar{\mathbf{U}}\bar{\mathbf{F}} = -\bar{\mathbf{U}}\bar{\mathbf{M}}\mathbf{X} + \bar{\mathbf{B}}^R \mathbf{\Omega}, \tag{3.11.22}$$

where

$$\boldsymbol{R} = [L_{i2\alpha} \exp(-ip\tau_\alpha h)], \qquad \boldsymbol{S} = [L_{i2\alpha} \exp(ip\tau_\alpha h)],$$
$$\mathbf{U} = [A_{k\alpha} \exp(ip\tau_\alpha h)], \qquad \mathbf{V} = [A_{k\alpha} \exp(-ip\tau_\alpha h)],$$
$$\boldsymbol{\Theta} = [\theta_j], \qquad \boldsymbol{\Omega} = [\omega_j], \qquad \mathbf{X} = [\chi_j],$$
$$\mathbf{E} = [E_j], \qquad \mathbf{F} = [F_j], \qquad \mathbf{M} = (M_{\alpha j}),$$
$$\mathbf{B}^L = [B_{kj}^L], \qquad \mathbf{B}^R = [B_{kj}^R].$$

Elimination of $\boldsymbol{\Theta}$ and $\boldsymbol{\Omega}$ from (3.11.19)–(3.11.22) yields

$$[\mathbf{S} - (\mathbf{B}^L)^{-1}\mathbf{V}]\mathbf{E} + [\bar{\mathbf{R}} - (\bar{\mathbf{B}}^L)^{-1}\bar{\mathbf{V}}]\bar{\mathbf{F}} = [-\mathbf{SM} + (\mathbf{B}^L)^{-1}\mathbf{UM}]\mathbf{X}, \qquad (3.11.23)$$

$$[\mathbf{S} - (\bar{\mathbf{B}}^R)^{-1}\mathbf{V}]\mathbf{E} + [\bar{\mathbf{R}} - (\bar{\mathbf{B}}^R)^{-1}\bar{\mathbf{U}}]\bar{\mathbf{F}} = [-\bar{\mathbf{S}}\bar{\mathbf{M}} + (\bar{\mathbf{B}}^R)^{-1}\bar{\mathbf{U}}\bar{\mathbf{M}}]\mathbf{X}. \qquad (3.11.24)$$

Elimination of $\bar{\mathbf{F}}$ yields

$$\mathbf{E} = \mathbf{QX}, \qquad (3.11.25)$$

where

$$\mathbf{Q} = \{[\bar{\mathbf{R}} - (\mathbf{B}^L)^{-1}\bar{\mathbf{V}}]^{-1}[\mathbf{S} - (\mathbf{B}^L)^{-1}\mathbf{U}] - [\bar{\mathbf{S}} - (\bar{\mathbf{B}}^R)^{-1}\bar{\mathbf{U}}]^{-1}[\mathbf{R} - (\bar{\mathbf{B}}^R)^{-1}\mathbf{V}]\}^{-1}$$
$$\times \{[\bar{\mathbf{R}} - (\mathbf{B}^L)^{-1}\bar{\mathbf{V}}]^{-1}[-\mathbf{SM} + (\mathbf{B}^L)^{-1}\mathbf{UM}] + \mathbf{M}\}. \qquad (3.11.26)$$

Similar expressions for $\mathbf{F}$, $\boldsymbol{\Theta}$ and $\boldsymbol{\Omega}$ (in terms of $\mathbf{X}$) may now be obtained by back substitution into (3.11.23), (3.11.19) and (3.11.20).

For simplicity, attention is now restricted to the case when $a_{ijkl}^L = a_{ijkl}^R$ so that $\mathbf{B}^L = \mathbf{B}^R$. Also, if the $p_i(x_1)$ are required to be even functions of x_1 then it is sufficient to take $\chi_i(p)$ in the form

$$\chi_i(p) = i \int_0^a r_i(t) J_0(pt)\, \mathrm{d}t \qquad (3.11.27)$$

where the $r_i(t)$ are real functions to be determined. With this choice of $\chi_i(p)$ the condition (3.11.10) is automatically satisfied. Also (3.11.23) and (3.11.24) may now be readily used to show that

$$\mathbf{E} = -\mathbf{F}. \qquad (3.11.28)$$

Use of (3.11.27) and (3.11.28) in (3.11.12) yields

$$\int_0^\infty \cos(px_1)p\, \mathrm{d}p \int_0^a r_j(t) J_0(pt)\, \mathrm{d}t$$

$$+ \int_0^\infty T_{jk}(p)\cos(px_1)p\, \mathrm{d}p \int_0^a r_k(t) J_0(pt)\, \mathrm{d}t$$

$$= -\pi p_j(x_1) \quad \text{for} \quad 0 < x_1 < a, \qquad (3.11.29)$$

where $T_{jk}(p)$ is the real matrix defined by

$$T_{jk}(p) = 2\mathscr{R} \sum_{\alpha} L_{j2\alpha} Q_{\alpha k}(p), \tag{3.11.30}$$

with the matrix $\mathbf{Q}$ given by (3.11.26). Proceeding as in Section 3.9 the equation (3.11.29) may be reduced to the form

$$r_j(t) + t \int_0^a K_{jk}(s, t) r_k(s)\, ds = -2t \int_0^t \frac{P_j(u)\, du}{(t^2 - u^2)^{1/2}}$$

$$\text{for} \quad 0 < t < a, \tag{3.11.31}$$

where

$$K_{jk}(s, t) = \int_0^\infty T_{jk}(p) J_0(ps) J_0(pt) p\, dp. \tag{3.11.32}$$

Equation (3.11.31) constitutes N simultaneous Fredholm integral equations for the $r_j(t)$, $j = 1, 2, \ldots, N$.

Note that if the prescribed $p_j(x_1)$ are not even functions of x_1 and/or $a_{ijkl}^L \neq a_{ijkl}^R$ then a solution to the problem in terms of two systems of N simultaneous Fredholm equations may be obtained by following the analysis of Section 3.9.

4

Some problems in elasticity and thermostatics

4.1 Prescribed temperature on the boundary of a half-space

Consider an anisotropic half-space $x_2 < 0$ with the temperature $T(x_1, x_2)$ prescribed on the boundary $x_2 = 0$. To find the temperature throughout the half-space it is necessary to obtain a solution to (2.3.2) which tends to zero as $|z| \to \infty$ and also satisfies the boundary condition

$$T(x_1, 0) = g(x_1), \tag{4.1.1}$$

where $g(x_1)$ is given. From Sections 2.3 and 3.2 it is apparent that the solution to this problem is

$$T = \theta_1(z_1) - \theta_1(\bar{z}_1) \quad \text{for} \quad x_2 < 0, \tag{4.1.2}$$

where

$$\theta_1(z) = -\frac{1}{2\pi i} \int_{-\infty}^{\infty} \frac{g(t)\, dt}{t - z} \tag{4.1.3}$$

and

$$z_1 = x_1 + \tau_1 x_2, \tag{4.1.4}$$

where τ_1 is given by (2.3.6). If

$$g(t) = \begin{cases} g_0 \text{(constant)} & \text{for} \quad a < x_1 < b, \\ 0 & \text{for} \quad x_1 < a \quad \text{and} \quad x_1 > b, \end{cases} \tag{4.1.5}$$

then

$$\theta_1(z) = -\frac{g_0}{2\pi i} \log \left(\frac{z - b}{z - a} \right). \tag{4.1.6}$$

Hence

$$T = -\frac{g_0}{2\pi i} \left\{ \log \left(\frac{z_1 - b}{z_1 - a} \right) - \log \left(\frac{\bar{z}_1 - b}{\bar{z}_1 - a} \right) \right\}. \tag{4.1.7}$$

Also, from (3.2.5), (2.3.7) and (4.1.6) it follows that, in this case

$$\psi_{1j} = -\frac{g_0}{2\pi i}\left\{(\lambda_{j1} + \tau_1\lambda_{j2})\left[\frac{b-a}{(z_1-b)(z_1-a)}\right]\right.$$
$$\left. - (\lambda_{j1} + \bar{\tau}_1\lambda_{j2})\left[\frac{b-a}{(\bar{z}_1-b)(\bar{z}_1-a)}\right]\right\}. \tag{4.1.8}$$

The heat flux $-P$ at a point across the surface with normal $\mathbf{n}$ is

$$P = \psi_{1j}n_j \tag{4.1.9}$$

with the ψ_{1j} given by (4.1.8). Hence the flux across the boundary $x_2 = 0$ is

$$-P = -\psi_{12} = \frac{g_0}{2\pi i}\cdot\left\{(\tau_1 - \bar{\tau}_1)\lambda_{22}\frac{b-a}{(x_1-b)(x_1-a)}\right\}$$
$$= \frac{g_0}{\pi}\left\{(\lambda_{11}\lambda_{22} - \lambda_{12}^2)^{1/2}\frac{b-a}{(x_1-b)(x_1-a)}\right\}. \tag{4.1.10}$$

Let

$$z_1 - b = r_1\exp(i\theta_1), \tag{4.1.11}$$
$$z_1 - a = r_2\exp(i\theta_2). \tag{4.1.12}$$

Equations (4.1.7) and (4.1.8) then yield

$$T = -\frac{g_0}{\pi}(\theta_1 - \theta_2) \tag{4.1.13}$$

$$\psi_{1j} = -\frac{g_0(b-a)}{2\pi i r_1 r_2}\{(\lambda_{j1} + \tau_1\lambda_{j2})\exp[-i(\theta_1 + \theta_2)]$$
$$- (\lambda_{j1} + \bar{\tau}_1\lambda_{j2})\exp[i(\theta_1 + \theta_2)]\}. \tag{4.1.14}$$

It is of interest to examine the case when $\lambda_{12} = 0$ and λ_{11} or λ_{22} is large. This case is of importance since in fibre-reinforced materials the thermal conductivity λ_{11} will often be large compared with λ_{22} or vice versa. With $\lambda_{12} = 0$ and $\tau_1 = i(\lambda_{11}/\lambda_{22})^{1/2}$ equation (4.1.14) becomes

$$\psi_{1j} = -\frac{g_0(b-a)}{\pi r_1 r_2}\{-\lambda_{j1}\sin(\theta_1 + \theta_2)$$
$$+ (\lambda_{11}/\lambda_{22})^{1/2}\lambda_{j2}\cos(\theta_1 + \theta_2)\}. \tag{4.1.15}$$

Hence the flux across the plane $x_2 = $ constant is

$$-\psi_{12} = \frac{g_0(b-a)}{\pi r_1 r_2}\{(\lambda_{11}\lambda_{22})^{1/2}\cos(\theta_1 + \theta_2)\}, \tag{4.1.16}$$

while the flux across the plane $x_1 = $ constant is

$$-\psi_{11} = -\frac{g_0(b-a)}{\pi r_1 r_2}\{\lambda_{11}\sin(\theta_1 + \theta_2)\}. \tag{4.1.17}$$

Now from (4.1.11) and (4.1.12)

$$r_1 \cos \theta_1 = x_1 - b, \qquad r_2 \cos \theta_2 = x_1 - a, \atop r_1 \sin \theta_1 = \varepsilon x_2, \qquad r_2 \sin \theta_2 = \varepsilon x_2, \quad\right\}} \tag{4.1.18}$$

where $\varepsilon = (\lambda_{11}/\lambda_{22})^{1/2}$. As $\varepsilon \to 0$ it follows from (4.1.18) that

$$\left. \begin{array}{lll} \theta_1 \to \pi, & \theta_2 \to \pi & \text{if } x_1 < a, \\ \theta_1 \to \pi, & \theta_2 \to 2\pi & \text{if } a < x_1 < b, \\ \theta_1 \to 2\pi, & \theta_2 \to 2\pi & \text{if } x_1 > b. \end{array} \right\} \tag{4.1.19}$$

Hence, from (4.1.13), (4.1.16) and (4.1.17)

$$T \to \begin{cases} -g_0 & \text{if } a < x_1 < b, \\ 0 & \text{if } x_1 < a \quad \text{or} \quad x_1 > b, \end{cases} \tag{4.1.20}$$

$$-\psi_{12} \to \begin{cases} -\dfrac{g_0(b-a)(\lambda_{11}\lambda_{22})^{1/2}}{(b-x_1)(x_1-a)} & \text{if } a < x_1 < b, \\[3mm] \dfrac{g_0(b-a)(\lambda_{11}\lambda_{22})^{1/2}}{\pi \, |x_1-b|\,|x_1-a|} & \text{if } x_1 < a \quad \text{or} \quad x_1 > b, \end{cases} \tag{4.1.21}$$

$$-\psi_{11} \to 0. \tag{4.1.22}$$

Thus if $\lambda_{12} = 0$ and the thermal conductivity λ_{22} is large compared with λ_{11} then the expressions for the temperature and fluxes $-\psi_{11}$ and $-\psi_{12}$ will be dominated by (4.1.20)–(4.1.22) until x_2 becomes significant compared to ε^{-1}.

As $\varepsilon \to \infty$ it follows from (4.1.18) that if the magnitude of x_2 is large compared with ε^{-1} then

$$\theta_1 \to 3\pi/2, \qquad \theta_2 \to 3\pi/2 \tag{4.1.23}$$

and hence, from (4.1.13) and (4.1.15),

$$T \to 0, \qquad -\psi_{11} \to 0, \qquad -\psi_{12} \to 0. \tag{4.1.24}$$

Thus if $\lambda_{12} = 0$ and the thermal conductivity λ_{11} is large compared with λ_{22} then the temperature and flux will be virtually zero outside of a narrow boundary layer in which x_2 is very small. Inside this layer the temperature and flux may be calculated by using (4.1.13) and (4.1.15) in conjunction with (4.1.18).

4.2 Prescribed heat flux on the boundary of a half-space

The heat flux $-P = -\psi_{12}$ is prescribed on the boundary $x_2 = 0$ of the anisotropic half-space $x_2 < 0$ as a function of x_1 only. Hence

$$\psi_{12}(x_1, 0) = p(x_1). \tag{4.2.1}$$

62   PROBLEMS IN ELASTICITY AND THERMOSTATICS

To find the temperature throughout the half-space it is necessary to
obtain a solution to (2.3.2) which satisfies the boundary condition (4.2.1).
From Sections 2.3 and 3.3 it follows that the solution to this problem is

$$T = (\lambda_{21} + \tau_1 \lambda_{22})^{-1} \chi_1(z_1) - (\lambda_{21} + \bar{\tau}_1 \lambda_{22})^{-1} \chi_1(\bar{z}_1), \tag{4.2.2}$$

$$\psi_{1j} = \left(\frac{\lambda_{j1} + \tau_1 \lambda_{j2}}{\lambda_{21} + \tau_1 \lambda_{22}} \right) \chi_1'(z_1) - \left(\frac{\lambda_{j1} + \bar{\tau}_1 \lambda_{j2}}{\lambda_{21} + \bar{\tau}_1 \lambda_{22}} \right) \chi_1'(\bar{z}_1), \tag{4.2.3}$$

with

$$\chi_1'(z_1) = -\frac{1}{2\pi i} \int_{-\infty}^{\infty} \frac{p(t)\,dt}{t-z}. \tag{4.2.4}$$

Suppose

$$p(x_1) = \begin{cases} p_0(\text{constant}) & \text{for} \quad a < x_1 < b, \\ 0 & \text{for} \quad x_1 < a \quad \text{and} \quad x_1 > b. \end{cases} \tag{4.2.5}$$

Then (4.2.4) yields

$$\chi_1'(z) = -\frac{p_0}{2\pi i} \log \left[\frac{z-b}{z-a} \right]. \tag{4.2.6}$$

Hence

$$\chi_1(z) = -\frac{p_0}{2\pi i} \{ (z-b) \log (z-b) - (z-a) \log (z-a) \}. \tag{4.2.7}$$

Note that $\chi_1(z)$ does not tend to zero as $|z| \to \infty$. The reason for this is
that

$$\int_{-\infty}^{\infty} p(t)\,dt = p_0(b-a) \neq 0$$

so that the condition (1.1.6) is not satisfied. The temperature field given
by (4.2.2) and (4.2.7) cannot be expected to give acceptable results at
large distances from the origin. However, it can be expected to give a
reasonably accurate temperature field near to that part of $x_2 = 0$ in which
the flux is prescribed. Specifically, using (4.1.11) and (4.1.12) in (4.2.7),

$$T = \frac{p_0}{2\pi i} \{ (\lambda_{21} + \tau_1 \lambda_{22})^{-1} [r_1 e^{i\theta_1} \log r_1 + i r_1 e^{i\theta_1} \theta_1 - r_2 e^{i\theta_2} \log r_2 - i r_2 e^{i\theta_2} \theta_2]$$

$$- (\lambda_{21} + \bar{\tau}_1 \lambda_{22})^{-1} [r_1 e^{-i\theta_1} \log r_1 - i r_1 e^{-i\theta_1} \theta_1$$

$$- r_2 e^{-i\theta_2} \log r_2 + i r_2 e^{-i\theta_2} \theta_2] \}. \tag{4.2.8}$$

On $x_2 = 0$ in the region $a < x_1 < b$, $\theta_1 = \pi$ and $\theta_2 = 2\pi$, so

$$T = -\frac{p_0}{2\pi i} \{(\lambda_{21} + \tau_1 \lambda_{22})^{-1}[-r_1 \log r_1 - \pi i r_1 - r_2 \log r_2 - 2\pi i r_2]$$

$$- (\lambda_{21} + \bar{\tau}_1 \lambda_{22})^{-1}[-r_1 \log r_1 + \pi i r_1 - r_2 \log r_2 + 2\pi i r_2]\}. \qquad (4.2.9)$$

If $\lambda_{12} = 0$ then $\tau_1 = i(\lambda_{11}/\lambda_{22})^{1/2}$ and hence (4.2.9) becomes

$$T = -\frac{p_0}{\pi(\lambda_{11}\lambda_{22})^{1/2}} \{r_1 \log r_1 + r_2 \log r_2\}. \qquad (4.2.10)$$

This temperature distribution tends to zero if either $\lambda_{11} \to \infty$ or $\lambda_{22} \to \infty$. Since (4.2.10) only determines T to within a constant it follows that T is nearly constant if λ_{11} or λ_{22} is very large.

4.3 An anisotropic elastic half-space with specified boundary tractions

Consider an anisotropic elastic half-space $x_2 < 0$ with the tractions P_i prescribed on the boundary $x_2 = 0$. The displacement and stress fields throughout the half-space are required. The boundary condition on $x_2 = 0$ is

$$P_i(x_1, 0) = \sigma_{i2}(x_1, 0) = p_i(x_1), \qquad (4.3.1)$$

where the $p_i(x_1)$, $i = 1, 2, 3,$, are given functions of x_1. From Sections 2.4 and 3.3 it follows that the solution to this problem may be written

$$u_k = \sum_\alpha A_{k\alpha} M_{\alpha j} \chi_j(z_\alpha) - \sum_\alpha \bar{A}_{k\alpha} \bar{M}_{\alpha j} \chi_j(\bar{z}_\alpha), \qquad (4.3.2)$$

$$\sigma_{ij} = \sum_\alpha L_{ij\alpha} M_{\alpha j} \chi_j'(z_\alpha) - \sum_\alpha \bar{L}_{ij\alpha} \bar{M}_{\alpha j} \chi_j'(\bar{z}_\alpha), \qquad (4.3.3)$$

with $\chi_i'(z)$ given by

$$\chi_i'(z) = -\frac{1}{2\pi i} \int_{-\infty}^{\infty} \frac{p_i(t)\,dt}{t - z}. \qquad (4.3.4)$$

In (4.3.2) and (4.3.3) the $A_{k\alpha}$, $L_{ij\alpha}$, $M_{\alpha j}$ and $z_\alpha = x_1 + \tau_\alpha x_2$ are given by (2.4.10)–(2.4.12). Since the constants occurring in (4.3.2) and (4.3.3) involve the roots of a sextic it will, in general, be necessary to proceed numerically to obtain the displacement and stress. However, in certain cases it is possible to obtain results analytically. For example, if $p_1(x_1)$ and $p_2(x_1)$ are both zero and

$$p_3(x_1) = \begin{cases} p_0 (\text{constant}) & \text{for} \quad a < x_1 < b, \\ 0 & \text{for} \quad x_1 < a \quad \text{and} \quad x_1 > b, \end{cases} \qquad (4.3.5)$$

then $\chi_1(z)$ and $\chi_2(z)$ may be taken to be zero and

$$\chi_3'(z) = -\frac{p_0}{2\pi i} \log \left[\frac{z-b}{z-a} \right]. \tag{4.3.6}$$

Integrating, it follows that

$$\chi_3(z) = -\frac{p_0}{2\pi i} \{(z-b) \log (z-b) - (z-a) \log (z-a)\}. \tag{4.3.7}$$

Further, if the $x_3 = 0$ plane is a plane of elastic symmetry then, from Section 2.5, equations (2.5.30)–(2.5.36) it is apparent that (4.3.2) and (4.3.3) reduce to

$$u_3 = A_{31} M_{13} \chi_3(z_1) - \bar{A}_{31} \bar{M}_{13} \chi_3(\bar{z}_1), \tag{4.3.8}$$

$$\sigma_{3j} = L_{3j1} M_{13} \chi_3'(z_1) - \bar{L}_{3j1} \bar{M}_{13} \chi_3'(\bar{z}_1), \tag{4.3.9}$$

where $z_1 = x_1 + \tau_1 x_2$ and τ_1 is the root with positive real part of the quadratic

$$c_{3131} + \tau c_{3132} + \tau c_{3231} + \tau^2 c_{3232} = 0. \tag{4.3.10}$$

Now A_{31} is a non-trivial solution to the equation

$$(c_{3131} + \tau c_{3132} + \tau c_{3231} + \tau^2 c_{3232}) A_{31} = 0, \tag{4.3.11}$$

so that it is possible to choose $A_{31} = 1$ and then, from (2.4.12),

$$L_{3j1} = (c_{3j31} + \tau_1 c_{3j32}). \tag{4.3.12}$$

Hence

$$M_{13} = L_{321}^{-1} = (c_{3231} + \tau_1 c_{3232})^{-1}. \tag{4.3.13}$$

Thus, substituting in (4.3.8) and (4.3.9),

$$u_3 = -\frac{p_0}{2\pi i} \{(c_{3231} + \tau_1 c_{3232})^{-1}[(z_1 - b) \log (z_1 - b) - (z_1 - a) \log (z_1 - a)]$$

$$- (c_{3231} + \bar{\tau}_1 c_{3232})^{-1}[(\bar{z}_1 - b) \log (\bar{z}_1 - b) - (\bar{z}_1 - a) \log (\bar{z}_1 - a)]\}, \tag{4.3.14}$$

$$\sigma_{3j} = -\frac{p_0}{2\pi i} \left\{ \frac{c_{3j31} + \tau_1 c_{3j32}}{c_{3231} + \tau_1 c_{3232}} \log \left[\frac{z-b}{z-d} \right] - \frac{c_{3j31} + \bar{\tau}_1 c_{3j32}}{c_{3231} + \bar{\tau}_1 c_{3232}} \log \left[\frac{\bar{z}-b}{\bar{z}-a} \right] \right\}. \tag{4.3.15}$$

If either the $x_1 = 0$ or $x_2 = 0$ plane is a plane of elastic symmetry then $c_{3132} = c_{3231} = 0$ and (4.3.10) yields

$$\tau_1 = i(c_{3131}/c_{3232})^{1/2}. \tag{4.3.16}$$

Putting $z - b = r_1 \exp (i\theta_1)$ and $z - a = r_2 \exp (i\theta_2)$ it follows from (4.3.14) that the displacement u_3 under the contact region $a < x_1 < b$ is given by

$$u_3 = -\frac{p_0}{\pi(c_{3131}c_{3232})^{1/2}}\{r_1 \log r_1 + r_2 \log r_2\}. \tag{4.3.17}$$

Hence if either of the shear moduli c_{3131}, c_{3232} is large then the displacement under the loaded region is small.

4.4 Mixed thermostatic problems for a half-space

The heat flux $-P = -\psi_{12}$ is prescribed over a section of the boundary $x_2 = 0$ and the temperature on the remainder. Also the temperature is required to tend to zero as $|z| \to \infty$. Let

$$\psi_{12}(x_1, 0) = p(x_1) \quad \text{for} \quad a < x_1 < b, \tag{4.4.1}$$

$$T(x_1, 0) = g(x_1) \quad \text{for} \quad x_1 < a \quad \text{and} \quad x_1 > b. \tag{4.4.2}$$

To find the temperature throughout the half-space $x_2 < 0$ it is necessary to obtain a solution to (2.3.2) which satisfies (4.4.1) and (4.4.2). From Sections 2.3 and 3.4 it follows that the required solution may be written in the form

$$T = \theta(z_1) - \theta(\bar{z}_1), \tag{4.4.3}$$

where $z_1 = x_1 + \tau_1 x_2$ with τ_1 given by (2.3.6) and

$$\theta'(z) = -\frac{X(z)}{2\pi i}\int_{-\infty}^{a}\frac{g'(t)\,dt}{X(t)(t-z)} - \frac{X(z)}{2\pi i(\lambda_{21}+\bar{\tau}_1\lambda_{22})}\int_{a}^{b}\frac{p(t)\,dt}{X^+(t)(t-z)}$$
$$-\frac{X(z)}{2\pi i}\int_{b}^{\infty}\frac{g'(t)\,dt}{X(t)(t-z)}, \tag{4.4.4}$$

where

$$X(z) = (z-b)^{m-1}(z-a)^{-m}, \tag{4.4.5}$$

$$m = \frac{1}{2\pi i}\log\left[\frac{\lambda_{21}+\bar{\tau}_1\lambda_{22}}{\lambda_{21}+\tau_1\lambda_{22}}\right], \tag{4.4.6}$$

where the branch of $X(z)$ is selected so that $zX(z) \to 1$ as $|z| \to \infty$ and the argument of $(\lambda_{21}+\bar{\tau}_1\lambda_{22})/(\lambda_{21}+\tau_1\lambda_{22})$ is chosen to lie between 0 and 2π. If T is constant on $x_2 = 0$ for $x_1 < a$ and $x_1 > b$ then (4.4.4) reduces to

$$\theta'(z) = \frac{X(z)}{2\pi i(\lambda_{21}+\bar{\tau}_1\lambda_{22})}\int_{a}^{b}\frac{p(t)\,dt}{X^+(t)(t-z)}. \tag{4.4.7}$$

If $\lambda_{21} = 0$ then $\tau_1 = i(\lambda_{11}/\lambda_{22})^{1/2}$ and hence (4.4.7) becomes

$$\theta'(z) = \frac{X(z)}{2\pi(\lambda_{11}\lambda_{22})^{1/2}}\int_{a}^{b}\frac{p(t)\,dt}{X^+(t)(t-z)}, \tag{4.4.8}$$

where

$$X(z) = (z-b)^{-1/2}(z-a)^{-1/2}. \tag{4.4.9}$$

A second mixed boundary-value problem for the half-space concerns determining the temperature in a half-space when the conditions on the boundary $x_2 = 0$ of the half-space are

$$P(x_1, 0) = \psi_{12}(x_1, 0) = p(x_1) \quad \text{for} \quad x_1 < a \quad \text{and} \quad x_1 > b, \tag{4.4.10}$$

$$T(x_1, 0) = g(x_1) \quad \text{for} \quad a < x_1 < b, \tag{4.4.11}$$

where, again, $p(x_1)$ and $g(x_1)$ are known functions of x_1. Using the results in Sections 2.3 and 3.5 the solution to this problem may be written

$$T = (\lambda_{21} + \tau_1 \lambda_{22})^{-1} \chi(z_1) - (\lambda_{21} + \bar{\tau}_1 \lambda_{22})^{-1} \chi(\bar{z}_1), \tag{4.4.12}$$

$$\psi_{1j} = \left[\frac{\lambda_{j1} + \tau_1 \lambda_{j2}}{\lambda_{21} + \tau_1 \lambda_{22}} \right] \chi'(z_1) - \left[\frac{\lambda_{j1} + \bar{\tau}_1 \lambda_{j2}}{\lambda_{21} + \bar{\tau}_1 \lambda_{22}} \right] \chi'(\bar{z}_1), \tag{4.4.13}$$

with

$$\chi'(z) = -\frac{X(z)}{2\pi i} \int_{-\infty}^{a} \frac{p(t)\,dt}{X(t)(t-z)} - \frac{(\lambda_{21} + \bar{\tau}_1 \lambda_{22})X(z)}{2\pi i} \int_{a}^{b} \frac{g'(t)\,dt}{X^+(t)(t-z)}$$

$$- \frac{X(z)}{2\pi i} \int_{b}^{\infty} \frac{p(t)\,dt}{X(t)(t-z)} + KX(z), \tag{4.4.14}$$

where

$$X(z) = (z-b)^{m-1}(z-a)^{-m}, \tag{4.4.15}$$

$$m = \frac{1}{2\pi i} \log \left[\frac{\lambda_{21} + \tau_1 \lambda_{22}}{\lambda_{21} + \bar{\tau}_1 \lambda_{22}} \right],$$

where, again, the branch of $X(z)$ is selected so that $zX(z) \to 1$ as $|z| \to \infty$ and the argument of $(\lambda_{21} + \tau_1 \lambda_{22})/(\lambda_{21} + \bar{\tau}_1 \lambda_{22})$ is chosen to lie between 0 and 2π.

4.5 Mixed elastostatic problems for a half-space

Consider an anisotropic elastic material which occupies the region $x_2 < 0$ and suppose the boundary $x_2 = 0$ is indented by a rigid punch which is linked to the half-space. This problem has been considered by Clements [9]. The boundary conditions are

$$\sigma_{12} = \sigma_{22} = \sigma_{23} = 0 \quad \text{for} \quad x_1 < a \quad \text{and} \quad x_1 > b, \tag{4.5.1}$$

$$u_k = f_k(x_1) \quad \text{for} \quad a < x_1 < b \quad \text{and} \quad k = 1, 2, 3, \tag{4.5.2}$$

$$\int_{a}^{b} P_j(x_1, 0)\,dx_1 = \int_{a}^{b} \sigma_{j2}(x_1, 0)\,dx_1 = -P_j \quad \text{for} \quad j = 1, 2, 3. \tag{4.5.3}$$

Expressions for the displacement and stress throughout the half-space may be obtained by employing the results of Sections 2.4 and 3.5. Specifically, if the stress is required to tend to zero as $|z| \to \infty$ then the solution takes the form

$$u_k = \sum_{\alpha} A_{k\alpha} M_{\alpha j} \chi_j(z_\alpha) - \sum_{\alpha} \bar{A}_{k\alpha} \bar{M}_{\alpha j} \chi_j(\bar{z}_\alpha), \tag{4.5.4}$$

$$\sigma_{ij} = \sum_{\alpha} L_{ij\alpha} M_{\alpha k} \chi'_k(z_\alpha) - \sum_{\alpha} \bar{L}_{ij\alpha} \bar{M}_{\alpha k} \chi'_k(\bar{z}_\alpha), \tag{4.5.5}$$

where

$$\chi'_k(z) = \sum_{\beta} \left[\frac{T_{k\beta} R_{\beta j} X_\beta(z)}{2\pi i} \int_a^b \frac{-f'_j(t)\, dt}{X_\beta^+(t)(t-z)} + T_{k\beta} K_\beta X_\beta(z) \right], \tag{4.5.6}$$

where $A_{k\alpha}$, $L_{ij\alpha}$, $M_{\alpha k}$ and $z_1 = x_1 + \tau_\alpha x_2$ are given by (2.4.10)–(2.4.12) while $T_{k\beta}$, $R_{\beta j}$ and $X_\beta(z)$ are defined by (3.4.18) and (3.3.5)–(3.3.10) and the K_β, $\beta = 1, 2, 3$ are constants. Also

$$K_\beta = -\frac{S_{\beta j} P_j}{2\pi i}, \tag{4.5.7}$$

where $S_{\gamma j}$ is defined by (3.5.6).

A second mixed problem involves the case when the boundary $x_2 = 0$ of the half-space $x_2 < 0$ is subjected to the conditions

$$u_k = 0 \qquad \text{for} \quad x_1 < a,\ x_1 > b \quad \text{and} \quad k = 1, 2, 3, \tag{4.5.8}$$

$$\sigma_{j2} = p_j(x_1) \quad \text{for} \quad a < x_1 < b \qquad \text{and} \quad j = 1, 2, 3. \tag{4.5.9}$$

This problem is of interest in determining the stress and displacement in an anisotropic elastic half-space which is bonded to a rigid material at all points of the boundary $x_2 = 0$ except those in the strip $a < x_1 < b$, $-\infty < x_3 < \infty$. Expressions for the stress and displacement throughout the half-space may be obtained by employing the results of Sections 2.4 and 3.4. Specifically,

$$u_k = \sum_{\alpha} A_{k\alpha} N_{\alpha j} \theta_j(z_\alpha) - \sum_{\alpha} \bar{A}_{k\alpha} \bar{N}_{\alpha j} \theta_j(\bar{z}_\alpha), \tag{4.5.10}$$

$$\sigma_{ij} = \sum_{\alpha} L_{ij\alpha} N_{\alpha k} \theta'_k(z_\alpha) - \sum_{\alpha} \bar{L}_{ij\alpha} \bar{N}_{\alpha k} \theta'_k(\bar{z}_\alpha), \tag{4.5.11}$$

where

$$\theta'_k(z) = \sum_{\gamma} \left\{ -\frac{T_{k\gamma} R_{\gamma i} X_\gamma(z)}{2\pi i} \int_a^b \frac{p_i(t)\, dt}{X_\gamma^+(t)(t-z)} \right\}. \tag{4.5.12}$$

4.6   Temperature field round a crack

Consider an anisotropic material which contains a plane crack whose width is small compared with the dimensions of the material. For the purpose of determining the temperature field round the crack it is useful to suppose that the material is infinite in extent since the error involved in this assumption will, in general, be negligible. The crack is considered to occupy the region $a < x_1 < b$, $-\infty < x_3 < \infty$ in the plane $x_2 = 0$ and the heat flux is given over the crack faces. Further, the heat flux is equal and opposite on the two faces and hence the total flux over the crack is zero. The boundary conditions over the crack are

$$\psi_{12}(x_1, 0) = p(x_1) \quad \text{for} \quad a < x_1 < b, \tag{4.6.1}$$

where $p(x_1)$ is given. Hence a solution to (2.3.2) is required which tends to zero as $|z| \to \infty$. From Sections 2.3 and 3.7 it follows that the solution to this problem may be written in the form

$$T = (\lambda_{21} + \tau_1 \lambda_{22})^{-1} \Psi(z_1) + (\lambda_{21} + \bar{\tau}_1 \lambda_{22})^{-1} \bar{\Psi}(\bar{z}_1) \quad \text{for} \quad x_2 > 0, \tag{4.6.2}$$

$$\psi_{1j} = \left[\frac{\lambda_{j1} + \tau_1 \lambda_{j2}}{\lambda_{21} + \tau_1 \lambda_{22}}\right] \Psi'(z_1) + \left[\frac{\lambda_{j1} + \bar{\tau}_1 \lambda_{j2}}{\lambda_{21} + \bar{\tau}_1 \lambda_{22}}\right] \bar{\Psi}'(\bar{z}_1) \quad \text{for} \quad x_2 > 0, \tag{4.6.3}$$

$$T = (\lambda_{21} + \tau_1 \lambda_{22})^{-1} \Omega(z_1) + (\lambda_{21} + \bar{\tau}_1 \lambda_{22})^{-1} \bar{\Omega}(\bar{z}_1) \quad \text{for} \quad x_2 < 0, \tag{4.6.4}$$

$$\psi_{1j} = \left[\frac{\lambda_{j1} + \tau_1 \lambda_{j2}}{\lambda_{21} + \tau_1 \lambda_{22}}\right] \Omega'(z_1) + \left[\frac{\lambda_{j1} + \bar{\tau}_1 \lambda_{j2}}{\lambda_{21} + \bar{\tau}_1 \lambda_{22}}\right] \bar{\Omega}'(\bar{z}_1) \quad \text{for} \quad x_2 < 0, \tag{4.6.5}$$

where $z_1 = x_1 + \tau_1 x_2$ and τ_1 is given by (2.3.6). Also

$$\Psi'(z) = \bar{\Omega}'(z) \quad \text{for} \quad x_2 > 0, \tag{4.6.6}$$

$$\Omega'(z) = \bar{\Psi}'(z) \quad \text{for} \quad x_2 < 0, \tag{4.6.7}$$

$$\Psi'(z) = \frac{X(z)}{2\pi i} \int_a^b \frac{p(t)\,dt}{X^+(t)(t - z)}, \tag{4.6.8}$$

$$X(z) = (z - a)^{-1/2}(z - b)^{-1/2}, \tag{4.6.9}$$

where the branch of $X(z)$ is selected so that $zX(z) \to 0$ as $|z| \to \infty$. Note that (4.6.8) defines $\Psi'(z)$ for all z in the cut plane. If $p(t) = p_0$(constant) then (4.6.8) yields

$$\Psi'(z) = \tfrac{1}{2} p_0 \{1 - [z - \tfrac{1}{2}(a + b)] X(z)\}, \tag{4.6.10}$$

so that, integrating,

$$\Psi(z) = \tfrac{1}{2}p_0\{z - (z-a)^{1/2}(z-b)^{1/2}\}. \tag{4.6.11}$$

Hence

$$T = \tfrac{1}{2}p_0\left\{\frac{z_1 - (z_1-a)^{1/2}(z_1-b)^{1/2}}{\lambda_{21} + \tau_1\lambda_{22}} + \frac{\bar{z}_1 - (\bar{z}_1-a)^{1/2}(\bar{z}_1-b)^{1/2}}{\lambda_{21} + \bar{\tau}_1\lambda_{22}}\right\}, \tag{4.6.12}$$

$$\psi_{1j} = \tfrac{1}{2}p_0\left\{\left[\frac{\lambda_{j1} + \tau_1\lambda_{j2}}{\lambda_{21} + \tau_1\lambda_{22}}\right]\left[1 - \frac{z_1 - \tfrac{1}{2}(a+b)}{(z_1-a)^{1/2}(z_1-b)^{1/2}}\right]\right.$$
$$\left. + \left[\frac{\lambda_{j1} + \bar{\tau}_1\lambda_{j2}}{\lambda_{21} + \bar{\tau}_1\lambda_{22}}\right]\left[1 - \frac{\bar{z}_1 - \tfrac{1}{2}(a+b)}{(\bar{z}_1-a)^{1/2}(\bar{z}_1-b)^{1/2}}\right]\right\}. \tag{4.6.13}$$

These expressions hold for the whole plane cut along the x_1-axis from a to b.

4.7 Stress field round a crack in an anisotropic material

Consider an infinite anisotropic material which contains a crack in the region $a < x_1 < b$, $-\infty < x_3 < \infty$ in the plane $x_2 = 0$ and suppose the tractions P_i are specified over the crack faces. Equal and opposite tractions are required to act on the two sides of the crack so that the total force over the crack is zero. The boundary conditions on the crack are therefore

$$\sigma_{i2} = p_i(x_1) \quad \text{for} \quad a < x_1 < b, \tag{4.7.1}$$

where the $p_i(x_1)$, $i = 1, 2, 3$ are given. Hence, a solution to (2.4.4) is required which tends to zero as $|z| \to \infty$ and satisfies the boundary condition (4.7.1). From Sections 2.4 and 3.7 it is apparent that the required solution to this problem may be written in the form

$$u_k = \sum_\alpha A_{k\alpha} M_{\alpha j} \Psi_j(z_\alpha) + \sum_\alpha \bar{A}_{k\alpha} \bar{M}_{\alpha j} \bar{\Psi}_j(\bar{z}_\alpha) \quad \text{for} \quad x_2 > 0, \tag{4.7.2}$$

$$\sigma_{ij} = \sum_\alpha L_{ij\alpha} M_{\alpha k} \Psi'_k(z_\alpha) + \sum_\alpha \bar{L}_{ij\alpha} \bar{M}_{\alpha k} \bar{\Psi}'_k(\bar{z}_\alpha) \quad \text{for} \quad x_2 > 0, \tag{4.7.3}$$

$$u_k = \sum_\alpha A_{k\alpha} M_{\alpha j} \Omega_j(z_\alpha) + \sum_\alpha \bar{A}_{k\alpha} \bar{M}_{\alpha j} \bar{\Omega}_j(\bar{z}_\alpha) \quad \text{for} \quad x_2 < 0, \tag{4.7.4}$$

$$\sigma_{ij} = \sum_\alpha L_{ij\alpha} M_{\alpha k} \Omega'_k(z_\alpha) + \sum_\alpha \bar{L}_{ij\alpha} \bar{M}_{\alpha k} \bar{\Omega}'_k(\bar{z}_\alpha) \quad \text{for} \quad x_2 < 0, \tag{4.7.5}$$

where the constants are defined by (2.4.10)–(2.4.12) and

$$\Psi_j'(z) = \bar{\Omega}_j'(z) \quad \text{for} \quad x_2 < 0, \tag{4.7.6}$$

$$\Omega_j'(z) = \bar{\Psi}_j'(z) \quad \text{for} \quad x_2 < 0, \tag{4.7.7}$$

$$\Psi_j'(z) = \frac{X(z)}{2\pi i} \int_a^b \frac{p_j(t)\, dt}{X^+(t)(t-z)}, \tag{4.7.8}$$

$$X(z) = (z-a)^{-1/2}(z-b)^{-1/2}, \tag{4.7.9}$$

where the branch of $X(z)$ is selected so that $zX(z) \to 0$ as $|z| \to \infty$. Now from (4.6.8) it follows that

$$\Psi'(z) = \frac{X(z)}{2\pi} \int_a^b \frac{(t-a)^{1/2}(b-t)^{1/2}p(t)\, dt}{t-z} \tag{4.7.10}$$

and in view of this formula and (4.7.6) and (4.7.7) it follows that (4.7.2) and (4.7.3) together with (4.7.10) give the displacement and stress at all points of the cut plane.

If the cut lies between $x_1 = -1$ and $x_1 = 1$ then the formula for the displacement and stress may be readily put into the form first derived by Stroh [38]. Specifically,

$$u_k = -\frac{1}{2\pi} \sum_\alpha (A_{k\alpha}M_{\alpha j} + \bar{A}_{k\alpha}\bar{M}_{\alpha j}) \int_0^1 d\mu \int_{-1}^1 p_j(\mu\xi)\xi(1-\xi^2)^{-1/2}\, d\xi$$

$$+ \frac{1}{2\pi} \sum_\alpha A_{k\alpha}M_{\alpha j} \int_0^1 \frac{d\mu}{(z_\alpha^2 - \mu^2)^{1/2}} \int_{-1}^1 \frac{(\mu + z_\alpha\xi)p_j(\mu\xi)\, d\xi}{(1-\xi^2)^{1/2}}$$

$$+ \frac{1}{2\pi} \sum_\alpha \bar{A}_{k\alpha}\bar{M}_{\alpha j} \int_0^1 \frac{d\mu}{(\bar{z}_\alpha^2 - \mu^2)^{1/2}} \int_{-1}^1 \frac{(\mu + \bar{z}_\alpha\xi)p_j(\mu\xi)\, d\xi}{(1-\xi^2)^{1/2}}, \tag{4.7.11}$$

where the constant term ensures that the displacement will tend to zero as $|z| \to \infty$.

$$\sigma_{ij} = -\frac{1}{2\pi} \sum_\alpha \left\{ \frac{L_{ij\alpha}M_{\alpha k}}{(z_\alpha^2 - 1)^{1/2}} \int_{-1}^1 \frac{(1-\xi^2)^{1/2}p_k(\xi)\, d\xi}{(z_\alpha - \xi)} \right.$$

$$\left. + \frac{\bar{L}_{ij\alpha}\bar{M}_{\alpha k}}{(\bar{z}_\alpha^2 - 1)^{1/2}} \int_{-1}^1 \frac{(1-\xi^2)^{1/2}p_k(\xi)\, d\xi}{(\bar{z}_\alpha - \xi)} \right\}. \tag{4.7.12}$$

In particular, in the important case when the applied tractions are uniform $p_i(x_1) = -p_i$(constant) and (4.7.12) yields

$$\sigma_{ij} = -\tfrac{1}{2}p_k \left\{ \sum_\alpha \{L_{ij\alpha}M_{\alpha k}[1 - z_\alpha(z_\alpha^2 - 1)^{-1/2}] + \bar{L}_{ij\alpha}\bar{M}_{\alpha k}[1 - \bar{z}_\alpha(\bar{z}_\alpha^2 - 1)^{-1/2}]\} \right\}. \tag{4.7.13}$$

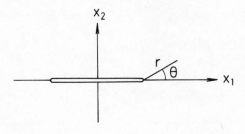

Fig. 4.7.1

Following the analysis of Stroh [38] it is possible to use (4.7.11) and (4.7.12) to examine the stress field round the crack tip and also to calculate the crack energy.

At points near the crack tip at $z = 1$ it is convenient to take $z_\alpha = 1 + \zeta_\alpha$ where $|\zeta_\alpha|$ is small. Let (Fig. 4.7.1)

$$\zeta_\alpha = r(\cos \theta + \tau_\alpha \sin \theta). \tag{4.7.14}$$

Now for small $|\zeta_\alpha|$

$$\frac{1}{\pi} \int_{-1}^{1} \frac{(1 - \xi^2)^{1/2} p_j(\xi)\, d\xi}{(z_\alpha - \xi)} \simeq \frac{1}{\pi} \int_{-1}^{1} \left[\frac{1 + \xi}{1 - \xi}\right]^{1/2} p_j(\xi)\, d\xi = -T_j, \text{ say.} \tag{4.7.15}$$

Hence, near the crack tip at $x_1 = 1$,

$$\sigma_{ij} = \tfrac{1}{2} \sum_\alpha \{L_{ij\alpha} M_{\alpha j}(2\zeta_\alpha)^{-1/2} + \bar{L}_{ij\alpha} \bar{M}_{\alpha j}(2\bar{\zeta}_\alpha)^{-1/2}\} T_j. \tag{4.7.16}$$

In the plane of the crack $\theta = 0$ and $\zeta_\alpha = r$ so that

$$\sigma_{i2} = (2r)^{-1/2} T_i. \tag{4.7.17}$$

Hence the tractions $P_i = \sigma_{i2}$ on the plane of the crack near the crack tip are independent of the elastic constants. Note that if the crack is of length $2a$ then the r in (4.7.17) should be replaced by r/a.

The energy of the crack is given by

$$U = -\tfrac{1}{2} \int_{-1}^{1} p_k(x_1) \Delta u_k\, dx_1, \tag{4.7.18}$$

where Δu_k is the relative displacement of the two sides of the crack. From (4.7.11) it follows that Δu_k is given by

$$\Delta u_k = -\frac{2}{\pi} B_{kj} \int_{|x_1|}^{1} \frac{d\mu}{(\mu^2 - x_1^2)^{1/2}} \int_{-1}^{1} \frac{(\mu + x_1 \xi) p_j(\mu \xi)\, d\xi}{(1 - \xi^2)^{1/2}}. \tag{4.7.19}$$

Hence

$$U = \tfrac{1}{2}\pi B_{kj} \int_{-1}^{1} |\mu| \, T_j(\mu) T_k(\mu) \, d\mu, \qquad (4.7.20)$$

where

$$T_j(\mu) = -\pi^{-1} \int_{-\pi/2}^{\pi/2} (1 + \sin\theta) p_i(\mu \sin\theta) \, d\theta. \qquad (4.7.21)$$

When the tractions are uniform $p_j(x_1) = -p_j$(constant) so that (4.7.20) becomes

$$U = \tfrac{1}{2}\pi B_{kj} p_j p_k. \qquad (4.7.22)$$

If the crack is of length $2a$ then the energy is

$$U = \tfrac{1}{2}\pi B_{kj} p_j p_k a^2. \qquad (4.7.23)$$

4.8 A crack between dissimilar anisotropic media

Assume two dissimilar anisotropic materials occupy the regions $x_2 > 0$ and $x_2 < 0$ which will be denoted by L and R respectively. The materials are assumed to be bonded at all parts of the interface $x_2 = 0$ except those lying in the region $|x_1| \leqslant a$, $-\infty < x_3 < \infty$ where there is a crack which is opened by equal and opposite tractions on each side of the crack. It is required to find the stress distribution in the bonded material. This problem has been considered by Clements [6] and Willis [41].

If the stresses and displacements in the regions L and R are denoted by σ_{ij}^L, u_k^L and σ_{ij}^R, u_k^R respectively then the following conditions must be satisfied on $x_2 = 0$:

$$\sigma_{i2}^L = p_i(x_1) \quad \text{on} \quad x_2 = 0+ \quad \text{for} \quad |x_1| < a, \qquad (4.8.1)$$

$$\sigma_{i2}^R = p_i(x_1) \quad \text{on} \quad x_2 = 0- \quad \text{for} \quad |x_1| < a, \qquad (4.8.2)$$

$$u_k^L = u_k^R \quad \text{on} \quad x_2 = 0 \quad \text{for} \quad |x_1| > a, \qquad (4.8.3)$$

$$\sigma_{i2}^L = \sigma_{i2}^R \quad \text{on} \quad x_2 = 0 \quad \text{for} \quad |x_1| > a, \qquad (4.8.4)$$

where the $p_i(x_1)$ are the given tractions over the crack faces. From Sections 2.4 and 3.8 it follows that a solution to the governing equation (2.4.4) which is zero at infinity and which satisfies the boundary conditions (4.8.1)–(4.8.4) may be written in the form

$$u_k^L = \sum_\alpha A_{k\alpha}^L M_{\alpha j}^L \Psi_j(z_\alpha) + \sum_\alpha \bar{A}_{k\alpha}^L \bar{M}_{\alpha j}^L \bar{\Psi}_j(\bar{z}_\alpha) \quad \text{for} \quad z_\alpha \in L, \qquad (4.8.5)$$

$$\sigma_{ij}^L = \sum_\alpha L_{ij\alpha}^L M_{\alpha k}^L \Psi_j'(z_\alpha) + \sum_\alpha \bar{L}_{ij\alpha}^L \bar{M}_{\alpha k}^L \bar{\Psi}_k'(\bar{z}_\alpha) \quad \text{for} \quad z_\alpha \in L, \qquad (4.8.6)$$

$$u_k^R = \sum_\alpha A_{k\alpha}^R M_{\alpha j}^R \Omega_j(z_\alpha) + \sum_\alpha \bar{A}_{k\alpha}^R \bar{M}_{\alpha j}^R \bar{\Omega}_j(\bar{z}_\alpha) \quad \text{for} \quad z_\alpha \in R, \tag{4.8.7}$$

$$\sigma_{ij}^R = \sum_\alpha L_{ij\alpha}^R M_{\alpha k}^R \Omega_k'(z_\alpha) + \sum_\alpha \bar{L}_{ij\alpha}^R \bar{M}_{\alpha k}^R \bar{\Omega}_k'(\bar{z}_\alpha) \quad \text{for} \quad z_\alpha \in R, \tag{4.8.8}$$

where the constants are defined by (2.4.10)–(2.4.12) and

$$\Psi_j'(z) = \bar{\Omega}_j'(z) \quad \text{for} \quad z \in L, \tag{4.8.9}$$

$$\Omega_j'(z) = \bar{\Psi}_j'(z) \quad \text{for} \quad z \in R, \tag{4.8.10}$$

with

$$\Psi_i'(z) = D_{ik}\chi_k'(z) \qquad \text{for} \quad z \in L, \tag{4.8.11}$$

$$\Psi_i'(z) = -\bar{D}_{ik}\chi_k'(z) \quad \text{for} \quad z \in R, \tag{4.8.12}$$

$$\chi_k'(z) = \sum_\gamma \left\{ \frac{T_{k\gamma}X_\gamma(z)}{2\pi i} \int_a^b \frac{R_{\gamma i}p_i(t)\,dt}{X_\gamma^+(t)(t-z)} \right\}, \tag{4.8.13}$$

$$X_\gamma(z) = (z-a)^{-m}(z-b)^{m-1}, \tag{4.8.14}$$

$$m = \frac{1}{2\pi i}\log\lambda_\gamma, \tag{4.8.15}$$

where the branch of $X_\gamma(z)$ is selected so that $zX_\gamma(z) \to 1$ as $|z| \to \infty$ and the argument of λ_γ is chosen to lie between 0 and 2π. In (4.8.11)–(4.8.15) the matrix $[D_{ij}]$ is defined by

$$(B_{kj}^L - \bar{B}_{kj}^L)D_{jl} = \delta_{kl}, \tag{4.8.16}$$

while the matrix $[R_{i\gamma}]$ is given by

$$(\bar{D}_{ik} - \lambda_\gamma D_{ik})R_{i\gamma} = 0, \tag{4.8.17}$$

where the λ_γ are the three roots of

$$|\bar{D}_{ik} - \lambda D_{ik}| = 0. \tag{4.8.18}$$

Also, the matrix $[T_{k\gamma}]$ is defined by

$$\sum_\gamma T_{i\gamma}S_{\gamma j} = \delta_{ij}, \tag{4.8.19}$$

where

$$S_{\gamma j} = R_{i\gamma}D_{ij}. \tag{4.8.20}$$

It is of particular interest to determine the values of λ_1, λ_2 and λ_3 since, through (4.8.13)–(4.8.15), these will determine the nature of the singularity at the crack tip. Clements [6] has determined numerical values for the elements of the matrices occurring in the equations of this section together with numerical values for the λ_γ for the case of a crack between two particular transversely isotropic materials. If the x_1-axis is normal to the transverse planes then the matrices $[A_{k\alpha}]$, $[L_{i2\alpha}]$ and $[M_{\alpha j}]$ are given by (2.7.6), (2.7.7) and (2.7.9). For the case when the materials in $x_2 > 0$ has constants $A = 16.2$, $N = 9.2$, $F = 6.9$, $C = 18.1$, $L = 4.67$ and the material in $x_2 < 0$ has constants $A = 5.92$, $N = 2.57$, $F = 2.14$, $C = 6.14$, $L = 1.64$ (where if each of these numerical values is multiplied by 10^{11} the units for the constants are dynes/cm^2) Clements finds that

$$[A^L_{k\alpha}] = \begin{bmatrix} 0 & 1.55 & 0.61 \\ 0 & i & i \\ 1 & 0 & 0 \end{bmatrix}, \quad [L^L_{i2\alpha}] = \begin{bmatrix} 0 & 14.28i & 6.94i \\ 0 & -10.78 & -8.71 \\ 4.04i & 0 & 0 \end{bmatrix},$$

$$[M^L_{\alpha j}] = \begin{bmatrix} 0 & 0 & -0.25i \\ -0.18i & 0.14 & 0 \\ 0.22i & -0.29 & 0 \end{bmatrix},$$

$$[B^L_{kj}] = \begin{bmatrix} -0.13i & 0.04 & 0 \\ -0.04 & -0.15i & 0 \\ 0 & 0 & -0.25i \end{bmatrix},$$

$$[A^R_{k\alpha}] = \begin{bmatrix} 0 & 1.93 & 0.51 \\ 0 & i & i \\ 1 & 0 & 0 \end{bmatrix}, \quad [L^R_{i2\alpha}] = \begin{bmatrix} 0 & 6.16i & 2.23i \\ 0 & -4.33 & -3.13 \\ 1.66i & 0 & 0 \end{bmatrix},$$

$$[M^R_{\alpha j}] = \begin{bmatrix} 0 & 0 & -0.6i \\ -0.33i & 0.23 & 0 \\ 0.45i & -0.64 & 0 \end{bmatrix},$$

$$[B^R_{kj}] = \begin{bmatrix} -0.4i & 0.12 & 0 \\ -0.12 & -0.41i & 0 \\ 0 & 0 & -0.6i \end{bmatrix}.$$

Hence

$$[D_{ij}] = \begin{bmatrix} 1.92i & -0.28 & 0 \\ 0.28 & 1.82i & 0 \\ 0 & 0 & 1.18i \end{bmatrix}$$

so that the roots of (4.8.18) are

$$\lambda_1 = -1, \qquad \lambda_2 = -0.75, \qquad \lambda_3 = -1.33. \tag{4.8.21}$$

A suitable choice of $[R_{\gamma i}]$ is then

$$[R_{\gamma i}] = \begin{bmatrix} 0 & 0 & 1 \\ -1.05i & 1 & 0 \\ 1.05i & 1 & 0 \end{bmatrix},$$

and therefore

$$[S_{\gamma k}] = \begin{bmatrix} 0 & 0 & 1.18i \\ 2.29 & 2.07i & 0 \\ -1.73 & 1.56i & 0 \end{bmatrix},$$

$$[T_{k\beta}] = \begin{bmatrix} 0 & 0.22 & -0.29 \\ 0 & -0.24i & -0.32i \\ -0.85i & 0 & 0 \end{bmatrix}.$$

This completes the calculation of the constants. Once the applied tractions are known, numerical values of the stress in the bonded material may be calculated through (4.8.5)–(4.8.13).

It has been shown (see Salganik [32] and England [17]) that violent oscillations occur in the stress near a straight crack between two bonded isotropic materials. This phenomenon is accompanied by interpretation of the crack surfaces which is a physically impossible condition. The values taken by λ in (4.8.21), together with (4.8.13), indicates that a similar situation exists for the anisotropic materials considered in this section. However, the oscillatory behaviour of the stress is confined to a small region around the crack tip in which the material is beyond the elastic limit and hence the linear theory of elasticity would not apply. Hence, for all practical purposes, this irregularity in the local stress may be ignored. It is of interest to note that, for the particular case considered here, the antiplane applied shear stress $\sigma_{23} = p_3(x_1)$ does not contribute to the oscillatory nature of the stress.

4.9   Temperature field in a slab

Consider an anisotropic material occupying the region between the planes $x_2 = \pm h$ and suppose that on the boundaries $x_2 = \pm h$ either the temperature or the heat flux is specified. From Sections 2.3 and 3.9 it follows that the temperature and heat flux throughout the slab will be given by the expressions

$$T = \frac{1}{\pi} \mathcal{R} \left\{ \int_0^\infty [E(p) \exp(ipz_1) + F(p) \exp(-ipz_1)] \, dp + D \right\}, \qquad (4.9.1)$$

$$\psi_{1j} = \frac{1}{\pi} \mathcal{R} \left\{ (\lambda_{j1} + \tau_1 \lambda_{j2}) \int_0^\infty [E(p) \exp(ipz_1) - F(p) \exp(-ipz_1)] ip \, dp \right\},$$

$$(4.9.2)$$

where $z_1 = x_1 + \tau_1 x_2$ with T_1 defined by (2.3.6). The flux $-P_1$ across a plane with normal n_j is given in terms of ψ_{1j} by

$$P_1 = \psi_{1j} n_j. \qquad (4.9.3)$$

The functions $E(p)$ and $F(p)$ are determined from the equations

$$\frac{1}{\pi} \mathcal{R}\left\{ \int_0^\infty [\exp(ip\tau_1 h)E + \exp(ip\bar{\tau}_1 h)\bar{F}] \exp(ipx_1)\, dp + D \right\} = T(x_1, h),$$

$$(4.9.4)$$

$$\frac{1}{\pi} \mathcal{R} \int_0^\infty [(\lambda_{21} + \tau_1\lambda_{22})\exp(ip\tau_1 h)E + (\lambda_{21} + \bar{\tau}_1\lambda_{22})\exp(ip\bar{\tau}_1 h)\bar{F}]$$

$$\times ip \exp(ipx_1)\, dp = \psi_{i2}(x_1, h), \quad (4.9.5)$$

$$\frac{1}{\pi} \mathcal{R}\left\{ \int_0^\infty [\exp(-ip\tau_1 h)E + \exp(-ip\bar{\tau}_1 h)\bar{F}] \exp(ipx_1)\, dp + D \right\}$$

$$= T(x_{12} - h), \quad (4.9.6)$$

$$\frac{1}{\pi} \mathcal{R} \int_0^\infty [(\lambda_{21} + \tau_1\lambda_{22})\exp(-ip\tau_1 h)E + (\lambda_{21} + \bar{\tau}_1\lambda_{22})\exp(-ip\bar{\tau}_1 h)\bar{F}]$$

$$\times \exp(ipx_1)\, dp = \psi_{i2}(x_1, -h). \quad (4.9.7)$$

For example, if on the boundaries $x_2 = \pm h$ the temperature is required to satisfy the conditions

$$T(x_1, h) = \begin{cases} T_0 & \text{for} \quad |x_1| < a, \\ 0 & \text{for} \quad |x_1| > a, \end{cases} \tag{4.9.8}$$

$$T(x_1, -h) = \begin{cases} T_0 & \text{for} \quad |x_1| < a, \\ 0 & \text{for} \quad |x_1| > a, \end{cases} \tag{4.9.9}$$

then the constant D may be put equal to zero and hence from the inversion theorem for Fourier transforms it follows that

$$\exp(ip\tau_1 h)E + \exp(ip\bar{\tau}_1 h)\bar{F} = \int_{-\infty}^\infty T(\xi, h)\exp(-i\xi p)\, d\xi$$

$$= 2T_0 p^{-1} \sin(ap), \tag{4.9.10}$$

$$\exp(-ip\tau_1 h)E + \exp(-ip\bar{\tau}_1 h)\bar{F} = \int_{-\infty}^\infty T(\xi, -h)\exp(-i\xi p)\, d\xi$$

$$= 2T_0 p^{-1} \sin(ap). \tag{4.9.11}$$

Hence

$$E(p) = F(p) = \frac{2T_0 \sin(ap)[\exp(-ip\bar{\tau}_1 h) - \exp(ip\bar{\tau}_1 h)]}{p\{\exp[ip(\tau_1 - \bar{\tau}_1)h] - \exp[-ip(\tau_1 - \bar{\tau}_1)h]\}}. \tag{4.9.12}$$

The temperature and heat flux throughout the slab are now given by (4.9.1)–(4.9.3) and (4.9.12).

4.10   Deformations of an anisotropic elastic slab

In this section deformations of the anisotropic elastic slab lying between the planes $x_2 = \pm h$ are considered. On $x_2 = \pm h$ either the displacement or the tractions are specified. From Sections 2.4 and 3.9 it follows that the displacement and stress throughout the slab are given by the expressions

$$u_k = \frac{1}{\pi} \mathcal{R} \sum_\alpha A_{k\alpha} \left\{ \int_0^\infty [E_\alpha(p) \exp(ipz_\alpha) + F_\alpha(p) \exp(-ipz_\alpha)] \, dp + D_\alpha \right\},$$

(4.10.1)

$$\sigma_{ij} = \frac{1}{\pi} \mathcal{R} \sum_\alpha L_{ij\alpha} \left\{ \int_0^\infty [E_\alpha(p) \exp(ipz_\alpha) - F_\alpha(p) \exp(-ipz_\alpha)] ip \, dp \right\},$$

(4.10.2)

where the constants are defined by (2.4.10)–(2.4.12) and the $E_\alpha(p)$ and $F_\alpha(p)$ may be obtained by employing the equations

$$\frac{1}{\pi} \mathcal{R} \left\{ \int_0^\infty \sum_\alpha [U_{k\alpha}E_\alpha + \bar{V}_{k\alpha}\bar{F}_\alpha] \exp(ipx_1) \, dp + \sum_\alpha A_{k\alpha}D_\alpha \right\} = u_k(x_1, h),$$

(4.10.3)

$$\frac{1}{\pi} \mathcal{R} \int_0^\infty \sum_\alpha [S_{i\alpha}E_\alpha + \bar{R}_{i\alpha}\bar{F}_\alpha] ip \exp(ipx_1) \, dp = \sigma_{i2}(x_1, h),$$

(4.10.4)

$$\frac{1}{\pi} \mathcal{R} \left\{ \int_0^\infty \sum_\alpha [V_{k\alpha}E_\alpha + \bar{U}_{k\alpha}\bar{F}_\alpha] \exp(ipx_1) \, dp + \sum_\alpha A_{k\alpha}D_\alpha \right\} = u_k(x_1, -h),$$

(4.10.5)

$$\frac{1}{\pi} \mathcal{R} \int_0^\infty \sum_\alpha [R_{i\alpha}E_\alpha + \bar{S}_{i\alpha}\bar{F}_\alpha] ip \exp(ipx_1) \, dp = \sigma_{i2}(x_1, -h),$$

(4.10.6)

where

$$R_{i\alpha} = L_{i2\alpha} \exp(-ip\tau_\alpha h), \tag{4.10.7}$$

$$S_{i\alpha} = L_{i2\alpha} \exp(ip\tau_\alpha h), \tag{4.10.8}$$

$$U_{k\alpha} = A_{k\alpha} \exp(ip\tau_\alpha h), \tag{4.10.9}$$

$$V_{k\alpha} = A_{k\alpha} \exp(-ip\tau_\alpha h). \tag{4.10.10}$$

Suppose the slab is subject to the loading shown in Fig. 4.10.1. The boundary conditions on $x_2 = \pm h$ are

$$\left. \begin{array}{l} \sigma_{12}(x_1, h) = \sigma_{32}(x_1, h) = 0, \\ \sigma_{22}(x_1, h) = [H(x_1 + d) - H(x_1 - d)]N_0, \end{array} \right\} \tag{4.10.11}$$

$$\left. \begin{array}{l} \sigma_{12}(x_1, h) = \sigma_{32}(x_1, -h) = 0, \\ \sigma_{22}(x_1, -h) = [H(x_1 + 10d) - H(x_1 - 10d)]\dfrac{N_0}{10}, \end{array} \right\} \tag{4.10.12}$$

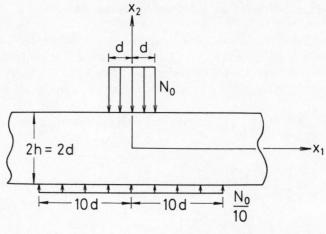

Fig. 4.10.1

where $H(x)$ is the Heaviside step function. Use of (4.10.4), (4.10.6), (4.10.11) and (4.10.12) together with the inversion theorem for Fourier transforms yields

$$\sum_{\alpha} [S_{j\alpha}E_{\alpha} + \bar{R}_{j\alpha}\bar{F}_{\alpha}] = -\frac{i\delta_{j2}N_0}{p} \int_{-\infty}^{\infty} \{H(\xi+d) - H(\xi-d)\} \exp(-ip\xi)\, d\xi$$

$$= \frac{2i\delta_{j2}N_0 \sin(pd)}{p^2}, \tag{4.10.13}$$

$$\sum_{\alpha} [R_{j\alpha}E_{\alpha} + \bar{S}_{j\alpha}\bar{F}_{\alpha}] = \frac{0.2i\delta_{j2}N_0 \sin(10\,pd)}{p^2}. \tag{4.10.14}$$

These equations may be readily solved to determine $E_{\alpha}(p)$ and $F_{\alpha}(p)$. Substitution into (4.10.1) and (4.10.2) then yields the displacement and stress throughout the material.

Tauchert [40] has used this formulation to consider deformations of an orthotropic elastic slab in which the principal directions of orthotropy are taken to be $0x_1'$, $0x_2'$ and $0x_3'$. A rotation of $-\theta$ about the $0x_3'$ axis yields the $0x_1x_2x_3$ frame so that (see Fig. 4.10.2)

$$x_i = a_{ij}x_j', \tag{4.10.15}$$

where

$$[a_{ij}] = \begin{bmatrix} \cos\theta & -\sin\theta & 0 \\ \sin\theta & \cos\theta & 0 \\ 0 & 0 & 1 \end{bmatrix}. \tag{4.10.16}$$

The elastic constants c_{ijkl} referred to the unprimed frame are related to

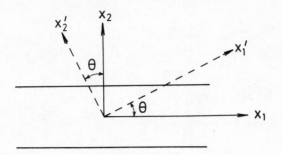

Fig. 4.10.2

those in the primed frame by

$$c_{ijkl} = a_{ip}a_{jr}a_{ks}a_{lt}c'_{prst}. \tag{4.10.17}$$

The numerical results presented by Tauchert are based on the ratios

$$\frac{c'_{2222}}{c'_{1111}} = 0.1, \qquad \frac{c'_{1122}}{c'_{1111}} = 0.03, \qquad \frac{c'_{1212}}{c'_{1111}} = 0.033.$$

These ratios were obtained from a particular boron-epoxy composite material. Using these ratios Tauchert calculated the stress and displacement in the slab with the loading given by (4.10.11) and (4.10.12). The procedure for obtaining numerical values of the stress and displacement is as follows.

Firstly the ratios c_{ijkl}/c'_{1111} are calculated by using (4.10.17). Then the τ_α, $A_{k\alpha}$ and $L_{ij\alpha}$ are obtained from (2.4.10)–(2.4.12). Equations (4.10.6)–(4.10.10) then yield $R_{i\alpha}$, $S_{i\alpha}$, $U_{k\alpha}$ and $V_{k\alpha}$ and then (4.10.13) and (4.10.14) may be used to find E_α and F_α. Use of a suitable quadrature formula in (4.10.1) and (4.10.2) then yields the displacement and stress at all points of the slab.

The stress $\sigma_{11}(0, x_2/h)$ at $\theta = 0$ and $\pi/2$ is given in Fig. 4.10.3. Also the deflection curve $u_2(x_1/h, 0)$ of the centreline of the slab is shown in Fig. 4.10.4 for various values of θ. It is interesting to note that when $\theta = 0$ the stress $\sigma_{11}(0, x_2/h)$ is much larger near the slab faces $x_2 = \pm h$ than when $\theta = \pi/2$. This is compatible with what might be expected on physical grounds since, when $\theta = 0$, the elastic stiffness c_{1111} is ten times as large as the stiffness c_{2222}.

A second problem considered by Tauchert concerns the case when the slab is fully constrained on $x_2 = -h$ and is subjected to a shear force on $x_2 = h$ (see Fig. 4.10.5). The boundary conditions are

$$\left. \begin{aligned} \sigma_{12}(x_1, h) &= [H(x_1 - d) - H(x_1 + d)]T_0, \\ \sigma_{22}(x_1, 0) &= \sigma_{32}(x_1, 0) = 0, \end{aligned} \right\} \tag{4.10.18}$$

$$u_k(x_1, -h) = 0 \quad \text{for} \quad k = 1, 2, 3. \tag{4.10.19}$$

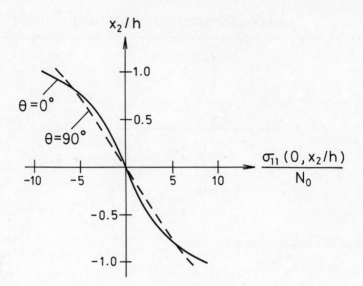

Fig. 4.10.3

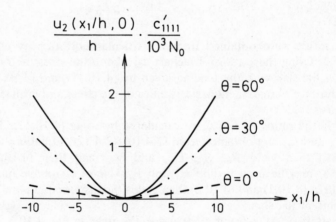

Fig. 4.10.4

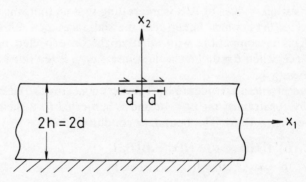

Fig. 4.10.5

80

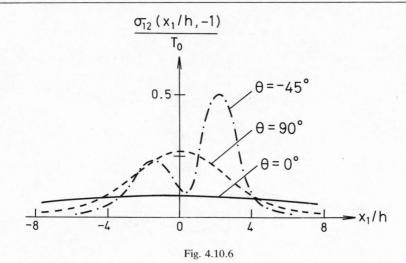

Fig. 4.10.6

Use of these conditions in (4.10.4) and (4.10.5) yields

$$\sum_{\alpha} [S_{j\alpha}E_{\alpha} + \bar{R}_{j\alpha}\bar{F}_{\alpha}] = -\frac{2i\delta_{ij}T_0 \sin{(pd)}}{p^2},$$

$$\sum_{\alpha} [V_{k\alpha}E_{\alpha} + \bar{U}_{k\alpha}\bar{F}_{\alpha}] = 0.$$

Proceeding in the manner described for the previous problem it is possible to obtain numerical values for the displacement and stress in a slab subjected to the boundary conditions (4.10.18) and (4.10.19). The shear stress σ_{12} and normal stress σ_{22} on the base line of the slab are shown in Figs 4.10.6 and 4.10.7. When $\theta = 0°$ (so that the reinforcing fibres of the composite are parallel to the slab faces) both σ_{12} and σ_{22} are relatively small. When the fibres are at a reasonably large angle to the slab faces the shear and normal stress on $x_2 = -h$ are considerably greater than when $\theta = 0°$.

4.11 Temperature field in a cracked slab

Suppose the anisotropic slab between the planes $x_2 = \pm h$ contains a crack in the region $x_2 = 0$, $-a < x_1 < a$, $-\infty < x_3 < \infty$ (Fig. 4.11.1). The temperature or heat flux is specified on the slab faces $x_2 = \pm h$ while on the crack the heat flux is zero. From Sections 2.3 and 3.10 it follows that the temperature and heat flux throughout the cracked slab will be given by

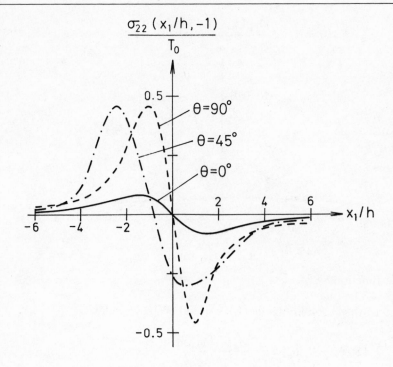

Fig. 4.10.7

the expressions

$$T = \frac{1}{\pi} \mathscr{R} \int_0^\infty \{[E(p) + (\lambda_{21} + \tau_1 \lambda_{22})^{-1} \chi(p)] \exp(ipz_1)$$
$$+ F(p) \exp(-ipz_1)\} \, dp \quad \text{for} \quad 0 < x_2 < h, \quad (4.11.1)$$

$$T = \frac{1}{\pi} \mathscr{R} \int_0^\infty \{E(p) \exp(ipz_1) + [F(p) + (\lambda_{21} + \tau_1 \lambda_{22})^{-1} \bar{\chi}(p)] \exp(-ipz_1)\} \, dp$$
$$\text{for} \quad -h < x_2 < 0, \quad (4.11.2)$$

$$\psi_{1j} = \frac{1}{\pi} \mathscr{R}(\lambda_{j1} + \tau_1 \lambda_{j2}) \int_0^\infty \{[E(p) + (\lambda_{21} + \tau_1 \lambda_{22})^{-1} \chi(p)] \exp(ipz_1)$$
$$- F(p) \exp(-ipz_1)\} ip \, dp \quad \text{for} \quad 0 < x_2 < h, \quad (4.11.3)$$

$$\psi_{1j} = \frac{1}{\pi} \mathscr{R}(\lambda_{j1} + \tau_1 \lambda_{j2}) \int_0^\infty \{E(p) \exp(ipz_1) - [F(p) + (\lambda_{21} + \tau_1 \lambda_{22})^{-1} \bar{\chi}(p)]$$
$$\times \exp(-ipz_1)\} ip \, dp \quad \text{for} \quad -h < x_2 < 0, \quad (4.11.4)$$

where $z_1 = x_1 + \tau_1 x_2$ with τ_1 given by (2.3.6) and

$$E(p) = Q(p)\chi(p) + H(p), \quad (4.11.5)$$
$$F(p) = Q(p)\bar{\chi}(p) + I(p). \quad (4.11.6)$$

In order to obtain explicit expressions for $Q(p)$, $H(p)$ and $I(p)$ suppose that ψ_{12} is given on $x_2 = \pm h$. Then use of the inversion theorem for Fourier transforms in, (4.11.3) and (4.11.4) yields

$$(\lambda_{21} + \tau_1 \lambda_{22})[E(p) + (\lambda_{21} + \tau_1 \lambda_{22})^{-1} \chi(p)] \exp{(ip\tau_1 h)}$$
$$+ (\lambda_{21} + \bar{\tau}_1 \lambda_{22})\bar{F}(p) \exp{(ip\bar{\tau}_1 h)} = \Theta(p), \quad (4.11.7)$$

$$(\lambda_{21} + \tau_1 \lambda_{22})E(p) \exp{(-ip\tau_1 h)} + (\lambda_{21} + \bar{\tau}_1 \lambda_{22})[\bar{F}(p)$$
$$+ (\lambda_{21} + \bar{\tau}_1 \lambda_{22})^{-1} \chi(p)] \exp{(-ip\bar{\tau}_1 h)} = \Omega(p), \quad (4.11.8)$$

where

$$\Theta(p) = -ip^{-1} \int_{-\infty}^{\infty} \psi_{12}(\xi, h) \exp{(-i\xi p)} \, \mathrm{d}\xi, \qquad (4.11.9)$$

$$\Omega(p) = -ip^{-1} \int_{-\infty}^{\infty} \psi_{12}(\xi, -h) \exp{(-i\xi p)} \, \mathrm{d}\xi. \qquad (4.11.10)$$

The functions $E(p)$ and $F(p)$ may be eliminated, in turn, from (4.11.7) and (4.11.8) to yield (4.11.5) and (4.11.6) with

$$Q(p) = (\lambda_{21} + \tau_1 \lambda_{22})^{-1} \{1 - \exp{[ip(\tau_1 - \bar{\tau}_1)h]}\} \{2 \sinh{[ip(\tau_1 - \bar{\tau}_1)h]}\}^{-1},$$
$$(4.11.11)$$

$$H(p) = (\lambda_{21} + \tau_1 \lambda_{22})^{-1} \{\Theta(p) \exp{(-ip\bar{\tau}_1 h)}$$
$$- \Omega(p) \exp{(ip\bar{\tau}_1 h)}\} \{2 \sinh{[ip(\tau_1 - \bar{\tau}_1)h]}\}^{-1}, \qquad (4.11.12)$$

$$I(p) = (\lambda_{21} + \tau_1 \lambda_{22})^{-1} \{\bar{\Omega}(p) \exp{(-ip\bar{\tau}_1 h)}$$
$$- \bar{\Theta}(p) \exp{(ip\bar{\tau}_1 h)}\} \{2 \sinh{[ip(\tau_1 - \bar{\tau}_1)h]}\}^{-1}. \qquad (4.11.13)$$

Similar expressions may be derived if the temperature is specified on both faces of the slab or, alternatively, if the temperature is specified on one face and the heat flux on the other.

Finally, the function $\chi(p)$ is given by

$$\chi(p) = \int_0^a s(t)J_1(pt) \, \mathrm{d}t + i \int_0^a r(t)J_0(pt) \, \mathrm{d}t, \qquad (4.11.14)$$

where

$$r(t) + t \int_0^a K^{(0)}(u, t)r(u) \, \mathrm{d}u = -\frac{t}{\pi} \int_{-t}^{t} \frac{P(u) \, \mathrm{d}u}{(t^2 - u^2)^{1/2}} \quad \text{for} \quad 0 < t < a,$$
$$(4.11.15)$$

$$s(t) + t \int_0^a K^{(1)}(u, t)s(u) \, \mathrm{d}u = -\frac{1}{\pi} \int_{-t}^{t} \frac{uP(u) \, \mathrm{d}u}{(t^2 - u^2)^{1/2}} \quad \text{for} \quad 0 < t < a$$
$$(4.11.16)$$

The kernels in these integral equations are given by

$$K^{(N)}(u, t) = \int_0^\infty T(p)J_N(pu)J_N(pt)p \, dp, \tag{4.11.17}$$

where

$$T(p) = 2\mathcal{R}\{(\lambda_{21} + \tau_1\lambda_{22})Q(p)\}, \tag{4.11.18}$$

while the function $P(x_1)$ in the integral equations is given by

$$P(x_1) = -\mathcal{R}\int_0^\infty [(\lambda_{21} + \tau_1\lambda_{22})H(p) + (\lambda_{21} + \bar{\tau}_1\lambda_{22})I(p)]ip \exp(ipx_1) \, dp.$$
$$\tag{4.11.19}$$

The procedure for determining the temperature and flux throughout the slab is as follows. Firstly, $\Theta(p)$ and $\Omega(p)$ are obtained from (4.11.9) and (4.11.10). $Q(p)$, $H(p)$ and $I(p)$ are then obtained from (4.11.11)–(4.11.13) so that (4.11.17)–(4.11.19) may be evaluated. The two Fredholm integral equations (4.11.15) and (4.11.16) may then be solved numerically. Next, equations (4.11.14), (4.11.5) and (4.11.6) yield $E(p)$ and $F(p)$ and then equations (4.11.1)–(4.11.4) yield the temperature and flux $P = \psi_{1j}n_j$.

4.12 Deformations of a cracked anisotropic slab

Consider the anisotropic elastic slab between the planes $x_2 = \pm h$ with a crack in the region $x_2 = 0$, $-a < x_1 < a$, $-\infty < x_3 < \infty$ (Fig. 4.11.1). Tractions are prescribed over both the crack faces and the boundary faces of the slab. From Sections 2.4 and 3.10 it follows that the displacement and stress throughout the cracked slab will be given by the expressions

$$u_k = \frac{1}{\pi}\mathcal{R}\sum_\alpha A_{k\alpha}\int_0^\infty \{[E_\alpha(p) + M_{\alpha i}\chi_i(p)]\exp(ipz_\alpha)$$
$$+ F_\alpha(p)\exp(-ipz_\alpha)\} \, dp \quad \text{for} \quad 0 < x_2 < h, \tag{4.12.1}$$

$$\sigma_{ij} = \frac{1}{\pi}\mathcal{R}\sum_\alpha L_{ij\alpha}\int_0^\infty \{[E_\alpha(p) + M_{\alpha k}\chi_k(p)]\exp(ipz_\alpha)$$
$$- F_\alpha(p)(-ipz_\alpha)\} ip \, dp \quad \text{for} \quad 0 < x_2 < h, \tag{4.12.2}$$

$$u_k = \frac{1}{\pi}\mathcal{R}\sum_\alpha A_{k\alpha}\int_0^\infty \{E_\alpha(p)\exp(ipz_\alpha)$$
$$+ [F_\alpha(p) + M_{\alpha i}\bar{\chi}_i(p)]\exp(-ipz_\alpha)\} \, dp \quad \text{for} \quad -h < x_2 < 0,$$
$$\tag{4.12.3}$$

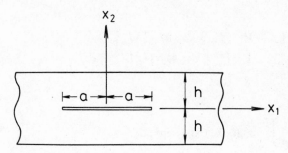

Fig. 4.11.1

$$\sigma_{ij} = \frac{1}{\pi} \mathscr{R} \sum_{\alpha} L_{ij\alpha} \int_0^\infty \{E_\alpha(p) \exp(ipz_\alpha)$$
$$- [F_\alpha(p) + M_{\alpha i}\bar{\chi}_i(p)] \exp(-ipz_\alpha)\} ip \, dp \quad \text{for} \quad -h < x_2 < 0,$$
$$(4.12.4)$$

where the $A_{k\alpha}$, $L_{ij\alpha}$ and $z_\alpha = x_1 + \tau_\alpha x_2$ are given by (2.4.10)–(2.4.12). On $x_2 = \pm h$ these equations yield

$$\frac{1}{\pi} \mathscr{R} \int_0^\infty \sum_\alpha [U_{k\alpha}(E_\alpha + M_{\alpha i}\chi_i) + \bar{V}_{k\alpha}\bar{F}_\alpha] \exp(ipx_1) \, dp = u_k(x_1, h),$$
$$(4.12.5)$$

$$\frac{1}{\pi} \mathscr{R} \int_0^\infty \sum_\alpha [S_{i\alpha}(E_\alpha + M_{\alpha k}\chi_k) + \bar{R}_{i\alpha}\bar{F}_\alpha] ip \exp(ipx_1) \, dp = \sigma_{i2}(x_1, h),$$
$$(4.12.6)$$

$$\frac{1}{\pi} \mathscr{R} \int_0^\infty \sum_\alpha [V_{k\alpha}E_\alpha + \bar{U}_{k\alpha}(\bar{F}_\alpha + \bar{M}_{\alpha i}\chi_i)] \exp(ipx_1) \, dp = u_k(x_1, -h),$$
$$(4.12.7)$$

$$\frac{1}{\pi} \mathscr{R} \int_0^\infty \sum_\alpha [R_{i\alpha}E_\alpha + \bar{S}_{i\alpha}(\bar{F}_\alpha + \bar{M}_{\alpha k}\chi_k)] ip \exp(ipx_1) \, dp = \sigma_{i2}(x_1, -h),$$
$$(4.12.8)$$

where

$$R_{i\alpha} = L_{i2\alpha} \exp(-ip\tau_\alpha h), \tag{4.12.9}$$
$$S_{i\alpha} = L_{i2\alpha} \exp(ip\tau_\alpha h), \tag{4.12.10}$$
$$U_{k\alpha} = A_{k\alpha} \exp(ip\tau_\alpha h), \tag{4.12.11}$$
$$V_{k\alpha} = A_{k\alpha} \exp(-ip\tau_\alpha h). \tag{4.12.12}$$

Suppose $\sigma_{i2}(x_1, h)$ and $\sigma_{i2}(x_1, -h)$ are given. Then (4.12.6) and (4.12.8) yield

$$\mathbf{SE} + \bar{\mathbf{R}}\bar{\mathbf{F}} = -\mathbf{SM}\boldsymbol{\chi} + \boldsymbol{\Theta}, \tag{4.12.13}$$
$$\mathbf{RE} + \bar{\mathbf{S}}\bar{\mathbf{F}} = -\bar{\mathbf{S}}\bar{\mathbf{M}}\boldsymbol{\chi} + \boldsymbol{\Omega}, \tag{4.12.14}$$

where

$$\mathbf{S} = [S_{i\alpha}], \qquad \mathbf{R} = [R_{i\alpha}], \qquad \mathbf{M} = [M_{\alpha i}],$$
$$\mathbf{E} = [E_\alpha], \qquad \mathbf{F} = [F_\alpha], \qquad \mathbf{\Theta} = [\Theta_i],$$
$$\mathbf{\Omega} = [\Omega_i], \qquad \mathbf{\chi} = [\chi_i]$$

with

$$\Theta_i = -\frac{i}{p} \int_{-\infty}^{\infty} \sigma_{i2}(\xi, h) \exp(-ip\xi) \, d\xi, \tag{4.12.15}$$

$$\Omega_i = -\frac{i}{p} \int_{-\infty}^{\infty} \sigma_{i2}(\xi, -h) \exp(-ip\xi) \, d\xi. \tag{4.12.16}$$

Elimination of $\mathbf{F}$ and $\mathbf{E}$, in turn, from (4.12.13) and (4.12.14) yields

$$E_\alpha = Q_{\alpha i}\chi_i + H_\alpha, \tag{4.12.17}$$

$$F_\alpha = Q_{\alpha i}\bar{\chi}_i + I_\alpha, \tag{4.12.18}$$

where

$$\mathbf{Q} = [Q_{\alpha i}] = \frac{\bar{\mathbf{M}} - \bar{\mathbf{R}}^{-1}\mathbf{SM}}{\bar{\mathbf{R}}^{-1}\mathbf{S} - \bar{\mathbf{S}}^{-1}\mathbf{R}}, \tag{4.12.19}$$

$$\mathbf{H} = [H_\alpha] = \frac{\bar{\mathbf{R}}^{-1}\mathbf{\Theta} - \bar{\mathbf{S}}^{-1}\mathbf{\Omega}}{\bar{\mathbf{R}}^{-1}\mathbf{S} - \bar{\mathbf{S}}^{-1}\mathbf{R}}, \tag{4.12.20}$$

$$\mathbf{I} = [I_\alpha] = \frac{\bar{\mathbf{R}}^{-1}\bar{\mathbf{\Omega}} - \bar{\mathbf{S}}^{-1}\bar{\mathbf{\Theta}}}{\bar{\mathbf{R}}^{-1}\mathbf{S} - \bar{\mathbf{S}}^{-1}\mathbf{R}}. \tag{4.12.21}$$

Similar forms for $\mathbf{Q}$, $\mathbf{H}$ and $\mathbf{I}$ are obtained if the displacement on $x_2 = \pm h$ is specified or, alternatively, if the displacement is specified on one of the faces of the slab and the tractions on the other.

The function $\chi_i(p)$ is given by

$$\chi_i(p) = \int_0^a s_i(t)J_1(pt) \, dt + i\int_0^a r_i(t)J_0(pt) \, dt, \tag{4.12.22}$$

where

$$r_j(t) + t\int_0^a K_{jk}^{(0)}(u, t)r_k(u) \, du = -\frac{t}{\pi} \int_{-t}^{t} \frac{P_j(u) \, du}{(t^2 - u^2)^{1/2}} \quad \text{for} \quad 0 < t < a, \tag{4.12.23}$$

$$s_j(t) + t\int_0^a K_{jk}^{(1)}(u, t)s_k(u) \, du = -\frac{1}{\pi} \int_{-t}^{t} \frac{uP_j(u) \, du}{(t^2 - u^2)^{1/2}} \quad \text{for} \quad 0 < t < a, \tag{4.12.24}$$

with

$$K_{jk}^{(N)}(u, t) = \int_0^\infty T_{jk}(p)J_N(pu)J_N(pt)p \, dp. \tag{4.12.25}$$

The function $T_{jk}(p)$ in (4.12.25) is defined by

$$T_{jk}(p) = 2\Re \sum_\alpha L_{j2\alpha} Q_{\alpha k}(p). \tag{4.12.26}$$

Also, in (4.12.23) and (4.12.24) $P_j(x_1)$ is given by

$$P_j(x_1) = \pi p_j(x) - \Re \int_0^\infty \sum_\alpha [L_{j2\alpha} H_\alpha(p) + \bar{L}_{j2\alpha} \bar{I}_\alpha(p)] ip \exp(ipx_1)\, dp, \tag{4.12.27}$$

where the given stress over the crack is $\sigma_{i2}(x_1, 0) = p_i(x_1)$.

The displacement and stress throughout the cracked slab may now be calculated as follows. Firstly, $\boldsymbol{\Theta}$ and $\boldsymbol{\Omega}$ are obtained from (4.12.15) and (4.12.16). $\mathbf{Q}$, $\mathbf{H}$ and $\mathbf{I}$ may then be obtained from (4.12.19)–(4.12.21) and hence $\tilde{P}_j(x_1)$, $T_{jk}(p)$ and $K_{jk}^{(N)}$ may be evaluated through (4.12.25)–(4.12.27). The integral equations (4.12.23) and (4.12.24) are then solved numerically. Equation (4.12.22) then yields $\chi_j(p)$ and hence $\mathbf{E}$ and $\mathbf{F}$ may be evaluated through (4.12.17) and (4.12.18). Finally, expressions (4.12.1)–(4.12.4) yield numerically values of the displacement and stress.

The energy of the crack is given by the integral

$$U = -\frac{1}{2} \int_{-a}^a p_k(x_1)\, \Delta u_k\, dx_1, \tag{4.12.28}$$

where Δu_k is obtained by subtracting (4.12.3) from (4.12.1). Hence

$$\Delta u_k = \frac{1}{\pi} \Re(B_{kj} - \bar{B}_{kj}) \int_0^\infty \chi_j(p) \exp(ipx_1)\, dp. \tag{4.12.29}$$

Clements [11] has calculated the crack energy for some particular traversely isotropic slabs with zero tractions on the boundaries $x_2 = \pm h$ and constant normal tractions over the crack. Thus the boundary conditions considered by Clements were

$$\sigma_{i2}(x_1, 0) = p_i(x_1) = -P_0\, \delta_{i2} \quad \text{for} \quad |x_1| < a, \tag{4.12.30}$$

$$\sigma_{j2}(x_1, h) = \sigma_{j2}(x_1, -h) = 0. \tag{4.12.31}$$

With these boundary conditions it is immediately apparent from (4.12.15) and (4.12.16) that $\boldsymbol{\Theta} = \boldsymbol{\Omega} = \mathbf{0}$ and hence, from (4.12.20) and (4.12.21), $\mathbf{H} = \mathbf{I} = \mathbf{0}$. Thus, from (4.12.27), $P_j(x_1) = -\pi P_0\, \delta_{j2}$ and it follows from (4.12.24) that $s_j(t) = 0$ so that (4.12.22) reduces to

$$\chi_j(p) = i \int_0^a r_j(t) J_0(pt)\, dt. \tag{4.12.32}$$

Substitution of (4.12.32) into (4.12.29) yields

$$\Delta u_k = \pi^{-1} i(B_{kj} - \bar{B}_{kj}) \int_0^a \frac{r_j(t)\, dt}{(t^2 - x_1^2)^{1/2}}. \tag{4.12.33}$$

Hence, in this case (4.12.28) gives the crack energy in the form

$$U = (2\pi)^{-1} i(B_{kj} - \bar{B}_{kj}) P_0 \, \delta_{k2} \int_0^a r_j(t) \, dt. \tag{4.12.34}$$

To obtain numerical values for the crack energy Clements puts $t = at'$. $s = as'$ and $p = p'/a$ so that (4.12.23), (4.12.25) and (4.12.34) become

$$[a^{-1} P_0^{-1} r_j(at')] + t' \int_0^1 K'_{jk}(s', t')[a^{-1} P_0^{-1} r_k(as')] \, ds' = \pi t' \, \delta_{i2}, \tag{4.12.35}$$

$$K'_{jk}(s', t') = \int_0^\infty T_{jk}(p'/a) J_0(p's') J_0(p't') p' \, dp', \tag{4.12.36}$$

$$U = \tfrac{1}{2} i(B_{2j} - \bar{B}_{2j}) P_0^2 a^2 \int_0^1 [a^{-1} P_0^{-1} r_j(at')] \, dt'. \tag{4.12.37}$$

As has been indicated in Section 2.7, traversely isotropic materials may be characterized by the five elastic constants A, N, F, C and L. The constants c_{ijkl} are obtained from these five constants and two angles α and θ through (2.7.1), (2.7.2) and (2.7.3).

Consider a material for which the elastic constants are $A = 5.96$, $N = 2.57$, $F = 2.14$, $C = 6.14$ and $L = 1.64$. If each of these constants is multiplied by 10^{11} then the units for the constants are dynes/cm^2. These are the constants for a crystal of magnesium although they are used merely for illustrative purposes. Using these constants and (4.12.35)–(4.12.37) it is possible to calculate the ratio U/U_0 (where U_0 denotes the energy of the corresponding crack in an infinite material). Some particular numerical values of this ratio for $\alpha = 0$, $\theta = \pi/2$ and various values of h/a are given in Table 4.12.1. The results indicate that, for the particular material under consideration, the difference between the energies of the crack in the strip and the crack in an infinite material is small provided the strip width is more than five times the crack length. If the strip width is less than three times the crack length then the difference in energies is appreciable.

Now consider materials with the same values of A, N, L and F as for magnesium but with various values of C. As C increases the extensibility of the material in a direction normal to the transverse plane decreases. As C becomes large, such a material will be a reasonable model of a fibre-reinforced material with almost inextensible straight fibres in an

Table 4.12.1 Variation of crack energy with strip width for $\alpha = 0$ and $\theta = \pi/2$.

h/a	1	2	3	4	5	10	20
U/U_0	2.19	1.31	1.14	1.08	1.05	1.01	1.00

Table 4.12.2 Variation of crack energy with C for $\alpha = 0$, $\theta = \pi/2$ and $h/a = 5$.

C	10	20	30	40	50	60	70	80	90	100
U/U_0	1.07	1.13	1.18	1.23	1.27	1.31	1.34	1.38	1.41	1.45

elastic matrix; the direction of the fibres being normal to the transverse planes (see, for example, Clements [8]). The variation of U/U_0 for various values of α, θ and C is shown in Tables 4.12.2, 4.12.3 and 4.12.4. It is interesting to note that a substantial increase in C only causes a marked change in U/U_0 in the case when $\alpha = 0$ and $\theta = \pi/2$. For these values of α and θ the crack lies in a transverse plane and the preferred direction is normal to the plane of the crack. If the preferred direction is at a substantial angle to the normal to the plane of the crack then, on the basis of the results in Tables 4.12.3 and 4.12.4, it is reasonable to conclude that an increase in inextensibility in the preferred direction has negligible effect on the crack energy.

Table 4.12.3 Variation of crack energy with C for $\alpha = \pi/6$, $\theta = \pi/4$ and $h/a = 5$.

C	10	20	30	40	50	60	70	80	90	100
U/U_0	1.04	1.04	1.05	1.05	1.05	1.05	1.05	1.05	1.05	1.05

Summarizing, the results show that for a material which exhibits normal extensibility in the preferred direction it is reasonable to use the energy for a crack in an infinite material as an approximation for the energy of a crack in a strip provided the strip width is greater than roughly five times the crack length. If, however, the material is almost inextensible in the preferred direction then even when $h/a \geqslant 5$ the difference between the energies of the two cracks may be appreciable.

Table 4.12.4 Variation of crack energy with C for $\alpha = \pi/3$, $\theta = \pi/4$ and $h/a = 5$.

C	10	20	30	40	50	60	70	80	90	100
U/U_0	1.05	1.05	1.06	1.06	1.06	1.07	1.07	1.07	1.08	1.08

4.13 Stress field in an anisotropic layered material with a crack

Suppose the regions $x_2 < -h$, $-h < x_2 < h$ and $x_2 > h$ are occupied by different anisotropic elastic materials and that in the plane $x_2 = 0$ there exists a crack in the region $-a < x_1 < a$, $-\infty < x_3 < \infty$ (Fig. 4.13.1). The crack is opened by equal and opposite tractions on each side of the crack. It is required to find the stress and displacement fields throughout the

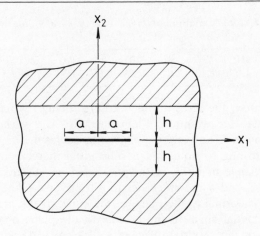

Fig. 4.13.1

material. If the materials in the region $x_2 < -h$ and $x_2 > h$ are required to be the same then, from Sections 2.4 and 3.10, it follows that the displacement and stress throughout the layered material will be given by

$$u_k^L = \frac{1}{\pi} \mathcal{R} \sum_\alpha A_{k\alpha}^L \int_0^\infty D_\alpha^L(p) \exp(ipz_\alpha^L) \, dp \quad \text{for} \quad x_2 > h, \tag{4.13.1}$$

$$\sigma_{ij}^L = \frac{1}{\pi} \mathcal{R} \sum_\alpha L_{ij\alpha}^L \int_0^\infty D_\alpha^L(p) \exp(ipz_\alpha^L) ip \, dp \quad \text{for} \quad x_2 > h, \tag{4.13.2}$$

$$u_k^R = \frac{1}{\pi} \mathcal{R} \sum_\alpha A_{k\alpha}^R \int_0^\infty D_\alpha^R(p) \exp(ipz_\alpha^R) \, dp \quad \text{for} \quad x_2 < -h, \tag{4.13.3}$$

$$\sigma_{ij}^R = \frac{1}{\pi} \mathcal{R} \sum_\alpha L_{ij\alpha}^R \int_0^\infty D_\alpha^R(p) \exp(ipz_\alpha^R) ip \, dp \quad \text{for} \quad x_2 < -h, \tag{4.13.4}$$

$$u_k = \frac{1}{\pi} \mathcal{R} \sum_\alpha A_{k\alpha} \int_0^\infty \{[E_\alpha(p) + M_{\alpha i}\chi_i(p)] \exp(ipz_\alpha)$$
$$- E_\alpha(p) \exp(-ipz_\alpha)\} \, dp \quad \text{for} \quad 0 < x_2 < h, \tag{4.13.5}$$

$$\sigma_{ij} = \frac{1}{\pi} \mathcal{R} \sum_\alpha L_{ij\alpha} \int_0^\infty \{[E_\alpha(p) + M_{\alpha i}\chi_i(p)] \exp(ipz_\alpha)$$
$$+ E_\alpha(p) \exp(-ipz_\alpha)\} ip \, dp \quad \text{for} \quad 0 < x_2 < h, \tag{4.13.6}$$

$$u_k = \frac{1}{\pi} \mathcal{R} \sum_\alpha A_{k\alpha} \int_0^\infty \{E_\alpha(p) \exp(ipz_\alpha) + [M_{\alpha i}\bar{\chi}_i(p)$$
$$- E_\alpha(p)] \exp(-ipz_\alpha)\} \, dp \quad \text{for} \quad -h < x_2 < 0, \tag{4.13.7}$$

$$\sigma_{ij} = \frac{1}{\pi} \mathcal{R} \sum_\alpha L_{ij\alpha} \int_0^\infty \{E_\alpha(p) \exp(ipz_\alpha) - [M_{\alpha i}\bar{\chi}_i(p)$$
$$- E_\alpha(p)] \exp(-ipz_\alpha)\} ip \, dp \quad \text{for} \quad -h < x_2 < 0, \tag{4.13.8}$$

where the $A_{k\alpha}$, $L_{ij\alpha}$ and $z_\alpha = x_1 + \tau_\alpha x_2$ are given by (2.4.10)–(2.4.12) and the superscripts L and R are used to denote the regions $x_2 > h$ and $x_2 < -h$ respectively. Let

$$\mathbf{R} = [L_{i2\alpha} \exp(-ip\tau_\alpha h)], \qquad \mathbf{S} = [L_{i2\alpha} \exp(ip\tau_\alpha h)],$$
$$\mathbf{U} = [A_{k\alpha} \exp(ip\tau_\alpha h)], \qquad \mathbf{V} = [A_{k\alpha} \exp(-ip\tau_\alpha h)],$$
$$\boldsymbol{\Theta} = [\theta_j], \qquad \boldsymbol{\Omega} = [\omega_j], \qquad \mathbf{X} = [\chi_j], \qquad \mathbf{E} = [E_j],$$
$$\mathbf{F} = [F_j], \qquad \mathbf{M} = [M_{\alpha j}], \qquad \mathbf{B}^L = [B_{kj}^L], \qquad \mathbf{B}^R = [B_{kj}^R],$$

with

$$D_\alpha^L(p) = M_{\alpha j}^L \exp(-ip\tau_\alpha^L h)\theta_j(p), \tag{4.13.9}$$
$$D_\alpha^R(p) = M_{\alpha j}^R \exp(ip\tau_\alpha^R h)\omega_j(p). \tag{4.13.10}$$

Then

$$\mathbf{E} = \mathbf{QX}, \tag{4.13.11}$$
$$\boldsymbol{\Theta} = \mathbf{SE} - \bar{\mathbf{R}}\bar{\mathbf{E}} + \mathbf{SMX}, \tag{4.13.12}$$
$$\boldsymbol{\Omega} = \mathbf{RE} - \bar{\mathbf{S}}\bar{\mathbf{E}} + \bar{\mathbf{S}}\bar{\mathbf{M}}\mathbf{X}, \tag{4.13.13}$$

where

$$\mathbf{Q} = \{[\bar{\mathbf{R}} - (\boldsymbol{\mathcal{B}}^L)^{-1}\bar{\mathbf{V}}]^{-1}[\mathbf{S} - (\mathbf{B}^L)^{-1}\mathbf{U}] - [\bar{\mathbf{S}} - (\bar{\mathbf{B}}^R)^{-1}\bar{\mathbf{U}}]^{-1}$$
$$\times [\mathbf{R} - (\mathbf{B}^R)^{-1}\mathbf{V}]^{-1}\}\{[\bar{\mathbf{R}} - (\mathbf{B}^L)^{-1}\bar{\mathbf{V}}]^{-1}[-\mathbf{SM} + (\mathbf{B}^L)^{-1}\mathbf{UM}] + \mathbf{M}\}. \tag{4.13.14}$$

The $\mathbf{X}$ in (4.13.11)–(4.13.13) is given by

$$\chi_j(p) = i\int_0^a r_j(t)J_0(pt)\,dt, \tag{4.13.15}$$

where

$$r_j(t) + t\int_0^a K_{jk}(s, t)r_k(s)\,ds = -2t\int_0^t \frac{p_j(u)\,du}{(t^2 - u^2)^{1/2}} \quad \text{for} \quad 0 < t < a, \tag{4.13.16}$$

with

$$K_{jk}(s, t) = \int_0^\infty T_{jk}(p)J_0(ps)J_0(pt)p\,dp, \tag{4.13.17}$$

$$T_{jk}(p) = 2\mathcal{R}\sum_\alpha L_{j2\alpha}Q_{\alpha k}(p). \tag{4.13.18}$$

Also, in (4.13.16), $\sigma_{j2}(x_1, 0) = p_j(x_1)$ are the given stresses over the crack faces.

Equations (4.13.1)–(4.13.18) may be used to calculate the displacement and stress throughout the layered material as follows. Firstly the roots τ_α

and the $A_{k\alpha}$, $L_{i2\alpha}$ and $M_{\alpha j}$ are calculated by using (2.4.10)–(2.4.12). The matrices $\mathbf{R}$, $\mathbf{S}$, $\mathbf{U}$, $\mathbf{V}$, $\mathbf{Q}$ and $\mathbf{T}$ may then be readily obtained and hence the integral equations (4.13.16) solved. Equation (4.13.15) then yields $\chi_i(p)$ so that (4.13.11)–(4.13.13) may be used to obtain $\mathbf{E}$, $\mathbf{\Theta}$ and $\mathbf{\Omega}$. The functions $D_\alpha^L(p)$ and $D_\alpha^R(p)$ are then calculated from (4.13.9) and (4.13.10). Finally, equations (4.13.1)–(4.13.8) yield values for the displacement and stress.

The energy of the crack is given by

$$U = -\frac{1}{2} \int_{-a}^{a} p_k(x_1)\, \Delta u_k \, dx_1, \tag{4.13.19}$$

where

$$\Delta u_k = \frac{1}{\pi}(B_{kj} - \bar{B}_{kj}) \int_0^\infty \chi_j(p) \exp(ipx_1)\, dp$$

$$= \pi^{-1} i (B_{kj} - \bar{B}_{kj}) \int_0^a \frac{r_j(t)\, dt}{(t^2 - x^2)^{1/2}}. \tag{4.13.20}$$

If the stresses $\sigma_{j2}(x_1, 0) = p_j(x_1)$ are such that $p_j(x_1) = -P_0\, \delta_{j2}$ then (4.13.19) yields

$$U = (2\pi)^{-1} i (B_{kj} - \bar{B}_{kj}) P_0\, \delta_{j2} \int_0^a r_j(t)\, dt. \tag{4.13.21}$$

To obtain values for the crack energy it is convenient to follow the analysis used by Clements [14] and put $t = at'$, $s = as'$ and $p = p'/a$ so that (4.13.16), (4.13.17) and (4.13.21) yield

$$[a^{-1} P_0^{-1} r_j(at')] + t' \int_0^1 K'_{jk}(s', t')[a^{-1} P_0^{-1} r_k(as')]\, ds' = \pi t'\, \delta_{j2}, \tag{4.13.22}$$

$$K'_{jk}(s', t') = \int_0^\infty T_{jk}(p'/a) J_0(p's') J_0(p't') p'\, dp', \tag{4.13.23}$$

$$U = \tfrac{1}{2} i (B_{2j} - \bar{B}_{2j}) P_0^2 a^2 \int_0^1 [a^{-1} P_0^{-1} r_j(at')]\, dt. \tag{4.13.24}$$

In the remainder of this section, the behaviour of (4.13.24) for various combinations of particular anisotropic materials is examined.

Consider transversely isotropic materials which may be characterized by the five constants A, N, F, C and L. The constants c_{ijkl} are obtained from these five constants and two angles α and θ through (2.7.1), (2.7.2) and (2.7.3). Two materials are needed for the composite under consideration; one for the region $-h < x_2 < h$ and one for the regions $x_2 < -h$ and $x_2 > h$. Let the constants for one of the materials (referred to subsequently as material I) be $A = 5.96$, $N = 2.57$, $F = 2.14$, $C = 6.14$ and

Table 4.13.1 Variation of crack energy with layer width for $\alpha = 0$ and $\theta = \pi/2$.

h/a	1	2	3	4	5	10	20
U/U_0 (Case I)	0.74	0.89	0.94	0.96	0.98	0.99	1.00
U/U_0 (Case II)	1.38	1.13	1.06	1.03	1.02	1.01	1.00

$L = 1.64$ while for the second material (material II) we use the constants $A = 16.2$, $N = 9.2$, $F = 6.9$, $C = 18.1$ and $L = 4.67$. If each of these numerical values is multiplied by 10^{11} then the units for the constants are dynes/cm^2. These are the values of the elastic constants for crystals of magnesium and titanium respectively although they are chosen here merely for illustrative purposes.

Using (4.13.22)–(4.13.24) the variation in U/U_0 (where U_0 denotes the energy of the corresponding crack in an infinite homogeneous material with the same constants as the material in $-h < x_2 < h$) may be calculated for $\alpha = 0$, $\theta = \pi/2$ (so that the crack lies in the transverse plane) and various values of h/a. The results, taken from Clements [14] are shown in Table 4.13.1. Case I in the table refers to a composite with material I in the region $-h < x_2 < h$ and material II in the regions $x_2 < -h$ and $x_2 > h$. The results in Table 4.13.1 show that, provided h/a is greater than five, the crack energy in the composite is practically the same as the energy of the corresponding crack in an infinite homogeneous material. If h/a is less than three then the difference between the materials in the half-space and the layer begins to appreciably influence the energy of the crack. When the material in the half-spaces is stronger than the material in the layer (Case I) then the energy of the crack in the composite decreases as h/a becomes small. Hence, according to the Griffith [21] theory of fracture, the crack becomes more stable as h/a becomes small since less energy is available for the formation of the new surface which is inherent in any crack extension. Conversely, if the material in the half-spaces is weaker than the material in the layer (Case II) then the energy of the crack increases as h/a becomes small so that the crack becomes less stable. Similar results to those shown in Table 4.13.1 were obtained for the angles $\alpha = \pi/6$, $\theta = \pi/4$ and $\alpha = \pi/3$, $\theta = \pi/4$. Hence it is reasonable to predict that this pattern would be repeated for all combinations of α and θ.

It is of interest to examine the case when the layer is "strongly anisotropic" since this case will be relevant to certain types of reinforced materials. In particular, it is of interest to consider a layer of material which is almost inextensible in a particular direction. Hence, let the layer $-h < x_2 < h$ consist of a material with the same values of A, N, F and L as for material I but with a larger value of C. As C becomes large, the inextensibility of the material in a direction normal to the transverse plane increases. The variation of U/U_0 with C for a layer of such a

Table 4.13.2 Variation of crack energy with C for $\alpha = 0$, $\theta = \pi/2$ and $h/a = 5$.

C	10	20	30	40	50	60	70	80	90	100
U/U_0	0.97	0.98	0.99	1.00	1.01	1.03	1.04	1.06	1.07	1.09
U/U_1	0.72	0.49	0.39	0.34	0.31	0.28	0.26	0.25	0.24	0.23

material sandwiched between two half-spaces of material II is shown in Tables 4.13.2, 4.13.3 and 4.13.4. Also the variation of U/U_1 (where U_1 is the value of the energy of the corresponding crack in an infinite region containing the homogeneous material I with $\alpha = 0$ and $\theta = \pi/2$) is shown in these three tables. It should be noted that the angles above each table refer only to the material in the region $-h < x_2 < h$. For the materials in $x_2 < -h$ and $x_2 > h$ the values of the angles were taken to be $\alpha = 0$ and $\theta = \pi/2$.

Table 4.13.3 Variation of crack energy with C for $\alpha = \pi/6$, $\theta = \pi/4$ and $h/a = 5$.

C	10	20	30	40	50	60	70	80	90	100
U/U_0	0.98	0.98	0.98	0.99	0.99	0.99	0.99	0.99	0.99	0.99
U/U_1	0.94	0.86	0.82	0.80	0.78	0.77	0.76	0.75	0.75	0.74

For the range of values of C considered, it is clear from the numerical results that there is little difference between the energies of the crack in the composite and the corresponding crack in an infinite homogeneous material with the same constants as the material in $-h < x_2 < h$. Thus, the stability of a crack in an almost inextensible layer of a composite would seem to be almost identical to the stability of the corresponding crack in an infinite material with the same constants as those of the layer. This observation is, of course, dependent on the crack being sufficiently removed from the material interfaces of the composite. If the crack is close to an interface between the layer and the surrounding weaker material then the results in Table 4.13.1 indicate that the crack would be less stable than the corresponding crack in an infinite material with the same constants as those of the layer.

The value of U/U_1 in Tables 4.13.2–4.13.4 show that an increase in C causes a decrease in the crack energy. This decrease in energy is most

Table 4.13.4 Variation of crack energy with C for $\alpha = \pi/3$, $\theta = \pi/4$ and $h/a = 5$.

C	10	20	30	40	50	60	70	80	90	100
U/U_0	0.98	0.99	0.99	0.99	0.99	0.99	1.00	1.00	1.00	1.00
U/U_1	0.88	0.74	0.69	0.65	0.63	0.61	0.60	0.59	0.59	0.58

marked when the crack lies in the transverse plane (so that $\alpha = 0$ and $\theta = \pi/2$).

4.14 The motion of a cylinder on an anisotropic half-space

A rigid smooth cylindrical body of given weight moves steadily over the surface of an anisotropic elastic material filling a half-space (Fig. 4.14.1). It is required to find the stress and velocity distribution throughout the half-space. The analysis of this section closely follows that of Clements [7].

Take Cartesian coordinates x_1, x_2, x_3 with x_2 vertical and consider the half-space $x_2 < 0$. Let the uniform velocity of the cylinder of radius R be $-V$. Write $X_1 = x_1 + Vt$, $X_2 = x_2$, $X_3 = x_3$ and let the area of contact between the cylinder and the elastic material be

$$-b < X_1 = x_1 + Vt < a, \qquad -\infty < X_3 < \infty,$$

where $a > 0$, $b > 0$ and $a + b \ll R$. Then the linear equations of elasticity are applicable and a solution to the problem is sought with the displacement and stress independent of X_3.

The stress-displacement relations and the equations of motion are

$$\sigma_{ij} = c_{ijkl}\frac{\partial u_k}{\partial x_l}$$

$$= c_{ijkl}\frac{\partial u_k}{\partial X_l}, \tag{4.14.1}$$

$$c_{ijkl}\frac{\partial^2 u_k}{\partial x_j\,\partial x_l} = \rho\frac{\partial^2 u_i}{\partial t^2} \tag{4.14.2}$$

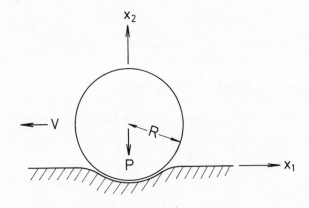

Fig. 4.14.1

or

$$d_{ijkl}\frac{\partial^2 u_k}{\partial X_j\, \partial X_l} = 0 \tag{4.14.3}$$

where ρ is the density and the d_{ijkl} are defined by (2.6.5). The analysis of the previous chapters is applicable to this case provided the $L_{ij\alpha}$ are still given by (2.4.12) and the $A_{k\alpha}$ are the solutions of the set of three homogeneous linear equations

$$(c_{i1k1} - \rho V^2\, \delta_{ik} + \tau_\alpha c_{i1k2} + \tau_\alpha c_{i2k1} + \tau_\alpha^2 c_{i2k2})A_{k\alpha} = 0, \tag{4.14.4}$$

where the τ_α, $\alpha = 1, 2, 3$ are the roots with the positive imaginary part of the sextic

$$|c_{i1k1} - \rho V^2\, \delta_{ik} + \tau c_{i1k2} + \tau c_{i2k1} + \tau^2 c_{i2k2}| = 0. \tag{4.14.5}$$

Only velocities V which lead to complex roots are admissible.

The boundary conditions on $x_2 = 0$ are

$$\sigma_{12}(X_1, 0) = \sigma_{23}(X_1, 0) = 0 \quad \text{for} \quad -\infty < X_1 < \infty, \tag{4.14.6}$$

$$\sigma_{22}(X_1, 0) = 0 \quad \text{for} \quad X_1 < -b \quad \text{and} \quad X_1 > a, \tag{4.14.7}$$

$$u_2(X_1, 0) = \beta X^2 - \gamma \quad \text{for} \quad -b < X_1 < a \tag{4.14.8}$$

with

$$\int_a^b \sigma_{22}(X_1, 0)\, dX_1 = -P, \tag{4.14.9}$$

where $\beta = 1/2R$, P is the given weight per unit length of the cylinder and γ, b and a are yet to be determined.

The problem is thus a special case of the more general problem considered in Section 3.6 (Problem I). The solution may be written in the form

$$u_k = \sum_\alpha A_{k\alpha} M_{\alpha j}\chi_j(z_\alpha) - \sum_\alpha \bar{A}_{k\alpha}\bar{M}_{\alpha j}\chi_j(\bar{z}_\alpha) \quad \text{for} \quad x_2 < 0, \tag{4.14.10}$$

$$\sigma_{ij} = \sum_\alpha L_{ij\alpha} M_{\alpha k}\chi_k'(z_\alpha) - \sum_\alpha \bar{L}_{ij\alpha}\bar{M}_{\alpha k}\chi_k'(\bar{z}_\alpha) \quad \text{for} \quad x_2 < 0, \tag{4.14.11}$$

where $z_\alpha = X_1 + \tau_\alpha X_2$, $\chi_1(z)$ and $\chi_3(z)$ are identically zero and

$$\chi_2'(z) = -\frac{X(z)}{2\pi i \bar{B}_{22}}\int_a^b \frac{2\beta t\, dt}{X^+(t)(t-z)} \tag{4.14.12}$$

with (see (3.6.14))

$$X(z) = (z-a)^{1/2}(z-b)^{1/2}. \tag{4.14.13}$$

To integrate (4.4.12) note that

$$\frac{1}{2\pi i} \oint \frac{\xi\,d\xi}{X(\xi)(\xi-z)} = \frac{1}{\pi i} \int_a^b \frac{t\,dt}{X^+(t)(t-z)}$$

where the integral on the left is taken clockwise along a simple contour surrounding the cut (a, b). Since the contour integral taken round a large circle in an anticlockwise direction is equal to one it follows, by residue theory,

$$\frac{1}{\pi i} \int_a^b \frac{t\,dt}{X^+(t)(t-z)} = \frac{z}{X(z)} - 1.$$

Substitution in (4.14.12) now yields

$$\chi_2'(z) = -\beta\bar{B}_{22}^{-1}[z - X(z)], \tag{4.14.14}$$

and for large $|z|$

$$\chi_2'(z) = \tfrac{1}{2}\beta\bar{B}_{22}^{-1}[a+b] + O(1/z). \tag{4.14.15}$$

In order to ensure that the stress vanishes at infinity it is necessary that $\chi_2'(z) \to 0$ as $|z| \to \infty$. Now from (4.14.15) it is clear that this condition will be satisfied provided

$$a = -b. \tag{4.14.16}$$

From (4.14.11), (4.14.14) and (4.14.16) an expression for the normal stress σ_{22} on the boundary $x_2 = 0$ may be obtained in the form

$$\sigma_{22} = -2i\beta\bar{B}_{22}^{-1}(b^2 - X_1^2)^{1/2} \quad \text{for} \quad |X_1| < b. \tag{4.14.17}$$

Hence condition (4.14.9) will be satisfied if

$$2i\beta\bar{B}_{22}^{-1} \int_{-b}^b (b^2 - X_1^2)^{1/2}\,dX_1 = P. \tag{4.14.18}$$

Upon integration this equation yields

$$i\beta\bar{B}_{22}^{-1}b^2\pi = P$$

or

$$b = \left\{\frac{P\bar{B}_{22}}{i\beta\pi}\right\}^{1/2}. \tag{4.14.19}$$

Equations (4.14.17) and (4.14.19) show that the stress under the cylinder is negative and b is real provided $\operatorname{Im} B_{22} < 0$. If $\operatorname{Im} B_{22} > 0$ then it is apparent that the stress under the cylinder is positive and b is imaginary – results which are incompatible with the requirement that the cylinder remains in contact with the half-space. The physical interpretation of the results is that the steady solution with finite stress exists for values of the velocity for which $\mathrm{Im}\,B_{22} < 0$ and the roots of the sextic (4.14.5) are complex. For such values of the velocity the solution shows that the indentation of the material is symmetric and corresponds to

$$-b(V) < X < b(V).$$

For the particular cases which have been examined by Clements [7] the roots of the sextic (4.14.5) are complex until the velocity reaches a certain value which will be denoted by V_0. Also $\mathrm{Im}\,B_{22}$ is initially (when $V = 0$) negative and decreases monotonically up to a certain velocity (V_1 say where $0 < V_1 < V_0$) which is such that $\mathrm{Im}\,B_{22} \to -\infty$ as $V \to V_1$. In the interval $V_1 < V < V_0$ $\mathrm{Im}\,B_{22}$ is positive so that no steady solution exists for V in this interval. Now since the displacement assumed on the surface was

$$u_2 = \beta X_1^2 - \gamma, \qquad -b < X_1 < b,$$

it follows that large b implies large deformations and the solution which was based on linearized theory is invalid. However, it is certainly safe to assume that the width (and depth) of indentation increases sharply as the velocity of the cylinder approaches V_1. It is of interest to note that a necessary condition for $\mathrm{Im}\,B_{22} \to -\infty$ as $V \to V_1$ is that $|L_{i2\alpha}| \to 0$ as $V \to V_1$. This condition may be established as follows. Now $B_{22} = \sum_\alpha A_{2\alpha} M_{\alpha 2}$ and recalling that, for each α, (4.14.4) determines only the ratios of the components, it follows that $A_{i\alpha}$ (and hence $L_{ij\alpha}$ and $M_{\alpha j}$) contains an arbitrary non-zero factor independent of i but which may depend on α. However, this factor does not appear in the product $A_{i\alpha} M_{\alpha j}$ so that $\sum_\alpha A_{i\alpha} M_{\alpha j}$ is unambiguously defined. Without loss of generality the non-zero $A_{i\alpha}$ (and hence $L_{ij\alpha}$) may be chosen to have finite modulus. It thus follows that $\mathrm{Im}\,B_{22} \to -\infty$ as $V \to V_1$ implies that $|L_{i2\alpha}| \to 0$ as $V \to V_1$.

This discussion of the behaviour of $\mathrm{Im}\,B_{22}$ has been limited to cases for which there exists supporting evidence. At the time of writing it has not proved possible to show that $\mathrm{Im}\,B_{22}$ behaves in this way for every anisotropic material and hence the possibility that $\mathrm{Im}\,B_{22}$ may behave in an alternative way for a particular class of anisotropic materials cannot be discounted.

The condition that $\mathrm{Im}\,B_{22}$ must be negative for the solution to be valid merits some comment. Now when $V = 0$ the solution is applicable to the static problem of a circular cylinder indenting an anisotropic half-space. It is thus reasonable to expect that for small enough V (and certainly for $V = 0$) the restriction that $\mathrm{Im}\,B_{22} < 0$ would be satisfied for every allowable choice of the constants c_{ijkl} (that is, for every choice which gives rise to a positive strain-energy density). To show that this is the case for general anisotropy would seem to be a difficult task but it is reasonably easy to consider the particular case of transversely isotropic materials when the x_3-axis is normal to the transverse plane (see Section 2.7). Use of (2.7.14) and (2.7.16) yields, in this case,

$$
\begin{aligned}
\operatorname{Im} B_{22} &= -\rho V^2 \{(A-N)^{1/2}(A-N-2\rho V^2)^{1/2} \\
&\quad - A^{1/2}(A-N-\rho V^2)^2(A-\rho V^2)^{-1/2}\}^{-1} \\
&= -2A/(A^2-N^2)+O(V^2).
\end{aligned}
\tag{4.14.20}
$$

Now two of the conditions which must be satisfied if the strain-energy density is to be positive are

$$c_{1111}>0, \qquad c_{1111}c_{2222}-c_{1122}^2>0$$

or in this case

$$A>0, \qquad A^2-N^2>0, \tag{4.14.21}$$

so that from (4.14.20) and (4.14.21) it is apparent that $\operatorname{Im} B_{22}<0$ for small enough V.

The formula (4.14.9) has been used by Clements [7] to evaluate $b[\beta/P]^{1/2}$ for two particular transversely isotropic materials. The material constants used were those for crystals of zinc and titanium. The constants for these two crystals are, in the notation of Section 2.7, $A=16.5$, $N=3.1$, $F=5$, $C=6.2$, $L=3.92$ and $A=16.2$, $N=9.2$, $F=6.9$, $C=18.1$, $L=4.67$ respectively. If each of these numerical values are multiplied by 10^{11} then the units for the constants are dynes/cm^2. The variation of $b[\beta/P]^{1/2}$ with ρV^2 is shown in Figs 4.14.2 and 4.14.3 with the angles

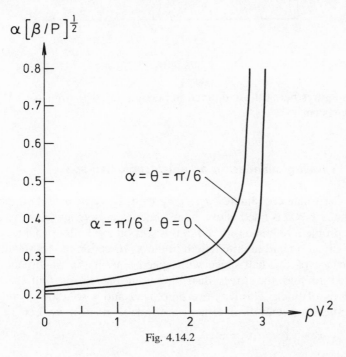

Fig. 4.14.2

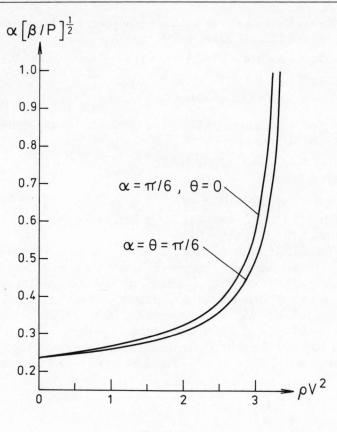

Fig. 4.14.3

on the figures being those defined in Section 2.7. The units for $\rho V^2 \times 10^{11}$ are dynes/cm^2.

4.15 A rolling cylinder on an anisotropic half-space

Take Cartesian coordinates x_1, x_2, x_3 with x_2 vertical and suppose the half-space $x_2 < 0$ is filled with a homogeneous anisotropic elastic material. In the problem to be considered in this section a cylinder of finite width in the x_1-direction and infinite length in the x_3-direction moves steadily over the surface of the half-space with speed $-V$ in the x_1-direction. It is required to find the stress distribution in the elastic half-space. The equations of linear elasticity are employed and a solution is sought for which the stress and displacement are independent of x_3. Let

$$X_1 = x_1 + Vt, \qquad X_2 = x_2, \qquad X_3 = x_3$$

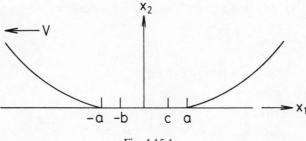

Fig. 4.15.1

and then the equations (4.14.1)–(4.14.5) are relevant to the problem of this section. The steadily rolling cylinder to be considered is in the form of an "inflated tyre" (Fig. 4.15.1). Such a cylinder will exert uniform normal pressure P and be instantaneously at rest over the central portion of the contact region. At both ends of the contact region, there will be an area in which the tyre will slip over the anisotropic elastic half-space $x_2 < 0$. Also it will be assumed that the tyre is in a state of limiting equilibrium in the axial direction so that the applied axial force will be equal to the product of the coefficient of friction and the force pressing the tyre into the half-space. The appropriate boundary conditions are

$$\sigma_{i2}(X_1, 0) = 0 \quad \text{for} \quad i = 1, 2, 3 \quad \text{and} \quad |X_1| > a, \tag{4.15.1}$$

$$\sigma_{22}(X_1, 0) = -P, \qquad \sigma_{23}(X_1, 0) = -\mu P \quad \text{for} \quad |X_1| < a, \tag{4.15.2}$$

$$\sigma_{12}(X_1, 0) = -\Lambda p \quad \text{for} \quad -a < X_1 < b, \tag{4.15.3}$$

$$\frac{\partial u_1(X_1, 0)}{\partial t} = 0 \quad \text{for} \quad -b < X_1 < c, \tag{4.15.4}$$

$$\sigma_{12}(X_1, 0) = \Lambda p \quad \text{for} \quad c < X_1 < a, \tag{4.15.5}$$

where μ equals plus or minus the coefficient of friction Λ, $2a$ is the width of the contact region which depends on the inflation of the tyre and the weight of the vehicle and b and c will be determined in terms of the load speed, the resultant applied force and the coefficient of friction. Also the stress must tend to zero at large distances from the region of contact.

The problem is thus a special case of the more general problem considered in Section 3.6 (Problem IV). The solution may be written in the form

$$u_k = \sum_\alpha A_{k\alpha} M_{\alpha j} \chi_j(z_\alpha) - \sum_\alpha \bar{A}_{k\alpha} \bar{M}_{\alpha j} \chi_j(\bar{z}_\alpha) \quad \text{for} \quad x_2 < 0, \tag{4.15.6}$$

$$\sigma_{ij} = \sum_\alpha L_{ij\alpha} M_{\alpha k} \chi'_k(z_\alpha) - \sum_\alpha \bar{L}_{ij\alpha} \bar{M}_{\alpha k} \chi'_k(\bar{z}_\alpha) \quad \text{for} \quad x_2 < 0 \tag{4.15.7}$$

where the $A_{k\alpha}$ are defined by (4.14.4) and, as usual, the $L_{ij\alpha}$ and $M_{\alpha k}$ are defined by (1.2.8) and (1.2.20), respectively. Also, $z_\alpha = X_1 + \tau_\alpha X_2$ and

$$
\begin{aligned}
\chi_2'(z) &= \frac{1}{2\pi i} \int_{-a}^{a} \frac{P\,dt}{t-z} \\
&= [P/2\pi i] \log[(z-a)/(z+a)] \\
&= \mu \chi_3'(z),
\end{aligned}
\tag{4.15.8}
$$

$$
\begin{aligned}
\chi_1'(z) &= \frac{X(z)}{2\pi i} \left[\int_{-a}^{-b} \frac{\Lambda P\,dt}{X(t)(t-z)} + \int_{-b}^{c} \frac{g(t)\,dt}{\bar{B}_{11}X^+(t)(t-z)} - \int_{c}^{a} \frac{\Lambda P\,dt}{X(t)(t-z)} \right] \\
&= \frac{\Lambda P}{\pi i} \log\left[\frac{(a-b)^{1/2}(z-c)^{1/2} + (a+c)^{1/2}(z+b)^{1/2}}{(b+c)^{1/2}(z+a)^{1/2}} \right] \\
&\quad + \frac{X(z)}{2\pi i} \int_{-b}^{c} \frac{g(t)\,dt}{\bar{B}_{11}X^+(t)(t-z)} \\
&\quad + \frac{\Lambda P}{\pi i} \log\left[\frac{(a-c)^{1/2}(z+b)^{1/2} + (a+b)^{1/2}(z-c)^{1/2}}{(b+c)^{1/2}(z-a)^{1/2}} \right],
\end{aligned}
\tag{4.15.9}
$$

where

$$
X(z) = (z+b)^{1/2}(z-c)^{1/2},
\tag{4.15.10}
$$

$$
\begin{aligned}
g(t) &= [P/2\pi i][B_{12} - \bar{B}_{12} + \mu(B_{13} - \bar{B}_{13})] \log[(a-t)/(a+t)] \\
&\quad - \tfrac{1}{2}P[(B_{12} + \bar{B}_{12}) + \mu(B_{13} + \bar{B}_{13})].
\end{aligned}
\tag{4.15.11}
$$

Since the stress must tend to zero at large distances from the contact region it is necessary that $\chi_1'(z) \to 0$ as $|z| \to \infty$ so that, from (4.15.9),

$$
\begin{aligned}
2\Lambda P \log &\{[(a-b)^{1/2} + (a+c)^{1/2}][(a-c)^{1/2} + (a+b)^{1/2}](b+c)^{-1}\} \\
&= \frac{1}{\bar{B}_{11}} \int_{-b}^{c} \frac{g(t)\,dt}{X^+(t)}.
\end{aligned}
\tag{4.15.12}
$$

If the resultant horizontal force on the cylinder is T then

$$
\int_{-a}^{a} \sigma_{12}(X_1, 0)\,dX = T.
\tag{4.15.13}
$$

Hence (3.6.54) yields

$$
\begin{aligned}
\int_{-a}^{a} \sigma_{12}(X_1, 0)\,dX = T = &-\Lambda P \int_{-a}^{b} \frac{(b-c+2t)\,dt}{2X(t)} \\
&- \int_{b}^{c} \frac{(b-c+2t)g(t)\,dt}{2\bar{B}_{11}X^+(t)} + \Lambda P \int_{c}^{a} \frac{(b-c+2t)\,dt}{2X(t)},
\end{aligned}
$$

or, after some elementary integration and some rearrangement,

$$
\begin{aligned}
b - c = &[(a-c)^{1/2}(a+b)^{1/2} + (a-b)^{1/2}(a+c)^{1/2}] \\
&\times \left\{ \frac{T}{2a P\Lambda} + \frac{1}{4a P\Lambda\bar{B}_{11}} \int_{-b}^{c} \frac{(b-c+2t)g(t)\,dt}{X^+(t)} \right\}.
\end{aligned}
\tag{4.15.14}
$$

Equations (4.15.12) and (4.15.14) determine b and c once T, P and Λ are known. It is clear that, in the general case, it will be necessary to determine b and c numerically but with the aid of an electronic computer this will not be particularly difficult. However, it is possible to obtain a certain amount of information from (4.15.12) and (4.15.14) without doing any numerical calculations. It is clear from (4.15.14) that if the product $P\Lambda$ is sufficiently large then the difference between b and c will be small. Hence the region of slip at the front of the tyre which we will denote by $S_F = (a-b)/a$ is nearly equal to the region of slip $S_R = (a-c)/a$ at the rear of the tyre. Also equation (4.15.14) shows that the effect of the driving or braking force T is to increase the zone of slip at one end of the contact region and decrease it at the other. Without doing any numerical work it is difficult to obtain much information, in the general case, about the influence which an applied axial force will have on the region of slip. However, when the $x_i = 0$ ($i = 1, 2,$ or 3) plane is a plane of elastic symmetry some simplification is obtained and we now examine these cases.

If the $x_1 = 0$ plane is a plane of elastic symmetry then, from (2.5.10) and (2.5.16), B_{12} and B_{13} have zero imaginary part so that, using (4.15.11), $g(t)$ reduces to

$$g(t) = -P(B_{12} + \mu B_{13}). \tag{4.15.15}$$

Hence, after some elementary integration (4.15.12) and (4.15.14) become

$$2\Lambda P \log\{[(a-b)^{1/2} + (a+c)^{1/2}][(a-c)^{1/2} + (a+b)^{1/2}](b+c)^{-1}\}$$
$$= P\pi i(B_{12} + \mu B_{13})/\bar{B}_{11}, \quad (4.15.16)$$

$$b - c = [(a-c)^{1/2}(a+b)^{1/2} + (a-b)^{1/2}(a+c)^{1/2}]T/2a P\Lambda. \tag{4.15.17}$$

It is immediately apparent from (4.15.17) that if the horizontal force T is zero then $b = c$ so that the region of slip is symmetrically placed about the centre $X_1 = 0$ of the contact region. Also, when $T = 0$, equation (4.15.16) may be written in the form

$$a/b = \cosh[\pi(B_{12} + \mu B_{13})/(2\Lambda |B_{11}|)], \tag{4.15.18}$$

where the properties of the B_{kj} outlined in Section 2.5 have been used to obtain this result. Hence the zones of slip are given by

$$S_F = S_R = (a-b)/a = 1 - \{\cosh[\pi(B_{12} + \mu B_{13})/(2\Lambda |B_{11}|)]\}^{-1}. \tag{4.15.19}$$

When the axial force is acting $\mu = \pm \Lambda$ so that it is clear from (4.15.19) that an axial force will either increase or decrease the width of the zones of slip.

If the $x_2 = 0$ plane is a plane of elastic symmetry then from Section 2.5 B_{12} has zero imaginary part and B_{13} has zero real part so that $g(t)$

reduces to

$$g(t) = [\mu B_{13} P/\pi i] \log [(a-t)/(a+t)] - PB_{12}.$$

Now if $\mu = 0$ so that the force in the axial direction is zero then $g(t) = -PB_{12}$ and (4.15.14) becomes

$$b - c = [(a-c)^{1/2}(a+b)^{1/2} + (a-b)^{1/2}(a+c)^{1/2}]T/2aP\Lambda,$$

and when $T = 0$ the regions of slips are symmetrically placed about $X_1 = 0$. If the axial force is non-zero so that $\mu = \pm \Lambda$ the integrals in (4.15.12) and (4.15.14) will not, in general, be zero so that even when $T = 0$ the regions of slip will not be symmetrically placed about the centre of the contact.

If the $x_3 = 0$ plane is a plane of elastic symmetry then $B_{13} = 0$. It follows from (4.15.11) that $g(t)$ is independent of the applied axial force. Hence equations (4.15.12) and (4.15.14) are independent of the axial force so it does not affect the zones of slip.

It thus appears that the width of the regions of slip will always depend upon the anisotropy of the material, the driving or braking force, the weight of the tyre and the coefficient of friction. However, they will only be affected by an applied axial force if the $x_3 = 0$ plane is not a plane of elastic symmetry. Now in isotropic materials the $x_3 = 0$ plane is clearly a plane of elastic symmetry and since the preceding analysis includes isotropy as a special case it follows that an axial force does not affect the zones of slip for an isotropic half-space.

The analysis of this section is restricted to velocities for which the sextic (4.14.5) has complex roots. As was noted in Section 4.14, this sextic does not have complex roots for all velocities but, for all the cases which have been tested so far, has complex roots in the interval going from $V = 0$ to a particular value $V = V_0$ say. Furthermore, within the interval $[0, V_0]$ there exists a velocity V_1 for which the matrix $[L_{i2\alpha}]$ is singular. As a result the moduli of the B_{kj} tend to infinity as $V \to V_1$ and hence the indentation made in the half-space by the cylinder tends to infinity as $V \to V_1$. Also, in the interval $[V_1, V_0]$ the solution to the boundary-value problem is inadmissible on physical grounds since it predicts the presence of the cylinder will cause a negative normal displacement on the surface of the half-space.

It is of interest to note that although the surface stresses σ_{12}, σ_{22}, σ_{32} are finite over the region of contact the remaining stresses have a logarithmic singularity at $z = \pm a$. This feature of the solution is also present in the related problem of an isotropic half-space subjected to a uniform shearing stress over an infinite strip (see, for example, Green and Zerna [20, page 266]). As a result the solution presented here can only be regarded as giving acceptable results for points not too close to $z = \pm a$.

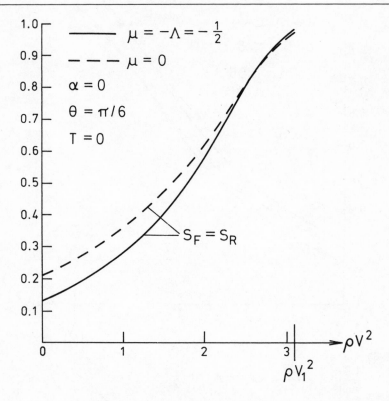

Fig. 4.15.2

Equations (4.15.12) and (4.15.14) have been used by Clements [10] to determine the region of slip for a particular transversely isotropic material. The material constants used were (in the notation of Section 2.7) $A = 16.2$, $N = 9.2$, $F = 6.9$, $C = 18.1$ and $L = 4.67$. If each of these numerical values are multiplied by 10^{11} then the units for these constants are dynes/cm^2. Using the two angles α and θ defined in Section 2.7 it is convenient to first consider the case when $\alpha = 0$ and $\theta = \pi/6$ so that the $x_1 = 0$ plane is a plane of elastic symmetry and equations (4.15.16) and (4.15.17) are applicable. For this case S_F and S_R are plotted against ρV^2 for $\mu = 0$ and $\mu = -\Lambda$ with zero driving force $T = 0$ and a driving force $T = Pa/4$ in Figs 4.15.2 and 4.15.3. The results for zero axial force ($\mu = 0$) are plotted on the same graphs for comparison. The units for $\rho V^2 \times 10^{11}$ are dynes/cm^2. It is apparent that, in this case, the effect of an axial force applied in a suitable direction is to reduce the width of the zones of slip. If the direction of the axial force is reversed then the zones of slip increase in width. It should be noted that, for the reasons mentioned in the preceding paragraphs, the results can only be considered to be reliable for values of ρV^2 not too near to ρV_1^2 where in this

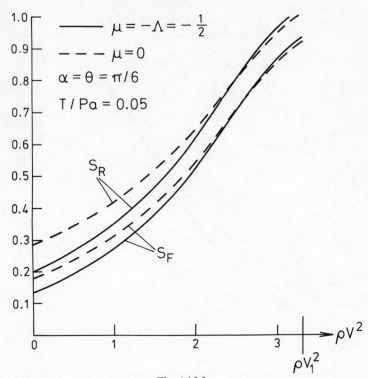

Fig. 4.15.3

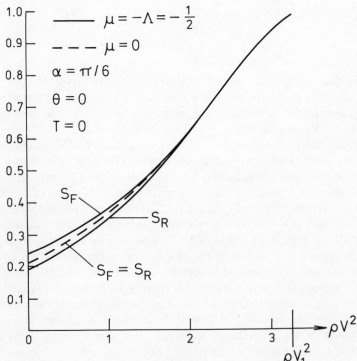

Fig. 4.15.4

106

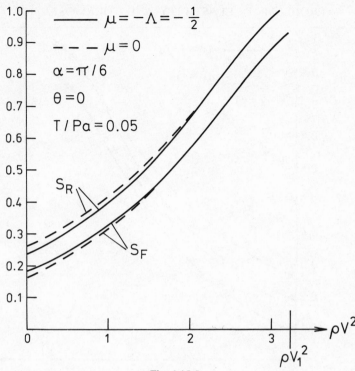

Fig. 4.15.5

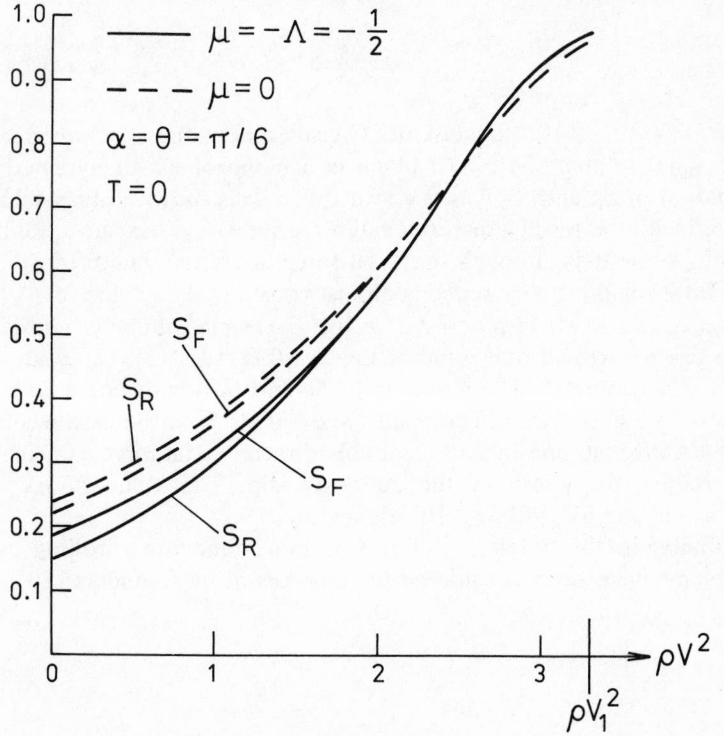

Fig. 4.15.6

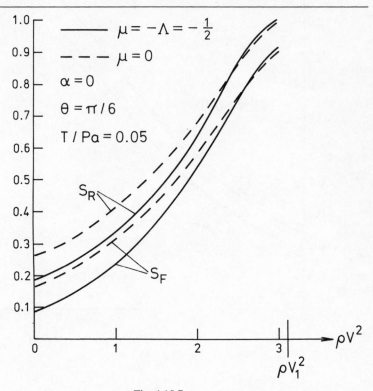

Fig. 4.15.7

case $\rho V_1^2 \approx 3.1 \times 10^{11}$ dynes/cm^2. Consider next the case when $\alpha = \pi/6$ and $\theta = 0$ so that the $x_2 = 0$ plane is a plane of elastic symmetry. The variation of S_F and S_R with ρV^2 for this case is shown in Figs 4.15.4 and 4.15.5 with the results for zero axial force shown on the same graphs. The results show that although the axial force alters the values of S_F and S_R the total region of slip remains almost constant. The value of ρV_1^2 when $\alpha = \pi/6$ and $\theta = 0$ is $\rho V_1^2 \approx 3.25 \times 10^{11}$ dynes/cm^2. Finally, consider the case $\alpha = \theta = \pi/6$ so that none of the $x_i = 0$ ($i = 1, 2, 3$) planes are planes of elastic symmetry. The variation of S_F and S_R for this case is shown in Figs 4.15.6 and 4.15.7. The results show that, as in the case when $\alpha = 0$ and $\theta = \pi/6$, the presence of a suitably directed axial force can considerably reduce the width of the zones of slip. The value of ρV_1^2 when $\alpha = \theta = \pi/6$ is $\rho V_1^2 \approx 3.35 \times 10^{11}$ dynes/cm^2.

Finally, in this section, it is noted that some other rolling cylinder problems have been considered in some detail by Clements [10].

5

Singular solutions to boundary-value problems

5.1 Introduction

In this chapter various solutions to the inhomogeneous system

$$a_{ijkl} \frac{\partial^2 \phi_k}{\partial x_j \, \partial x_l} = K_i \, \delta(\mathbf{x} - \mathbf{x}_0), \qquad \mathbf{x} = (x_1, x_2), \qquad \mathbf{x}_0 = (a, b)$$

$$\text{for} \quad i = 1, 2, \ldots, N, \quad (5.1.1)$$

are constructed where δ denotes the Dirac delta functional, the K_i are constants, and the a_{ijkl} satisfy the conditions outlined in Section 1.1. In Chapter 6 these solutions will be combined with a desired solution of the homogeneous system (1.1.1) through a reciprocal relation to yield boundary integral formulae for the required solution to (1.1.1).

5.2 Fundamental singular solution

To construct a solution to (5.1.1) consider functions

$$f_\alpha(z_\alpha) = \frac{1}{2\pi i} D_\alpha \log(z_\alpha - c_\alpha) \tag{5.2.1}$$

where $c_\alpha = a + \tau_\alpha b$ and the D_α are constants. Substitution of (5.2.1) into (1.2.4) yields

$$\phi_k = \frac{1}{2\pi i} \left\{ \sum_\alpha A_{k\alpha} D_\alpha \log(z_\alpha - c_\alpha) - \sum_\alpha \bar{A}_{k\alpha} \bar{D}_\alpha \log(\bar{z}_\alpha - \bar{c}_\alpha) \right\}. \tag{5.2.2}$$

Along any closed path encircling the point $\mathbf{x}_0$, ϕ_k as given by (5.2.2) changes by an amount

$$b_k = \sum (A_{k\alpha} D_\alpha + \bar{A}_{k\alpha} \bar{D}_\alpha). \tag{5.2.3}$$

109

This change in b_k will be zero if the D_α are defined by

$$D_\alpha = \tfrac{1}{2} i N_{\alpha j} d_j \qquad (5.2.4)$$

where the d_j are real constants and the $N_{\alpha j}$ are given by (1.2.12). Thus

$$\phi_k = \frac{1}{4\pi} \left\{ \sum_\alpha A_{k\alpha} N_{\alpha j} \log (z_\alpha - c_\alpha) + \sum_\alpha \bar{A}_{k\alpha} \bar{N}_{\alpha j} \log (\bar{z}_\alpha - \bar{c}_\alpha) \right\} d_j \qquad (5.2.5)$$

may be regarded as a possible solution to (5.1.1). The solution is valid everywhere with logarithmic singularity at the point (a, b) which is admissible according to familiar interpretations of the Dirac delta functional. Further, it is readily verified that (5.2.5) has the remaining necessary properties to satisfy (5.1.1) if the K_i in (5.1.1) are related to the d_r according to

$$K_i = -\tfrac{1}{2} i \sum_\alpha [L_{i2\alpha} N_{\alpha r} - \bar{L}_{i2\alpha} \bar{N}_{\alpha r}] d_r. \qquad (5.2.6)$$

5.3   First problem for a half-plane

A solution to (5.1.1) is required which is valid in the half-plane $x_2 < 0$. On the boundary $x_2 = 0$ of the half-plane the dependent variables are zero. That is

$$\phi_k (x_1, 0) = 0. \qquad (5.3.1)$$

To construct this solution consider the functions

$$f_\alpha (z_\alpha) = \frac{1}{4\pi} N_{\alpha j} d_j \log (z_\alpha - c_\alpha) + g_\alpha (z_\alpha) \quad \text{for} \quad \alpha = 1, 2, \ldots, N,$$

$$(5.3.2)$$

where the d_j are constants and the $g_\alpha (z_\alpha)$ are analytic in the half-plane $x_2 < 0$. Also, $\mathbf{x}_0$ is required to lie in $x_2 < 0$ so $b < 0$ and hence $c_\alpha = a + \tau_\alpha b$ lies in $x_2 < 0$. Substitution of (5.3.2) into (1.2.4) yields

$$\phi_k = \frac{1}{4\pi} \left\{ \sum_\alpha A_{k\alpha} N_{\alpha j} \log (z_\alpha - c_\alpha) + \sum_\alpha \bar{A}_{k\alpha} \bar{N}_{\alpha j} \log (\bar{z}_\alpha - \bar{c}_\alpha) \right\} d_j$$

$$+ \sum_\alpha A_{k\alpha} g_\alpha (z_\alpha) + \sum_\alpha \bar{A}_{k\alpha} \bar{g}_\alpha (\bar{z}_\alpha). \qquad (5.3.3)$$

Now since the functions $g_\alpha (z_\alpha)$ are required to be analytic in $x_2 < 0$ it follows from the previous section that (5.3.3) will be a solution to (5.1.1) if the constants d_j and K_i are related by the equation

$$K_i = -\tfrac{1}{2} i \sum_\alpha [L_{i2\alpha} N_{\alpha j} - \bar{L}_{i2\alpha} \bar{N}_{\alpha j}] d_j. \qquad (5.3.4)$$

Thus, with d_j given by (5.3.4), (5.3.3) represents a solution to (5.1.1) valid in the half-plane $x_2 < 0$. It remains to choose the analytic functions $g_\alpha(z_\alpha)$ so that the boundary conditions (5.3.1) are satisfied. Image considerations lead to the choice

$$g_\alpha(z_\alpha) = -\frac{1}{4\pi} \left\{ N_{\alpha q} \sum_\beta \bar{A}_{q\beta} \bar{N}_{\beta j} \log(z_\alpha - \bar{c}_\beta) \right\} d_j. \tag{5.3.5}$$

These functions are clearly analytic in $x_2 < 0$ since $\bar{c}_\beta$ lies in the upper half-plane $x_2 > 0$. Substitution of (5.3.5) into (5.3.3) yields

$$\phi_k = \frac{1}{4\pi} \left\{ \sum_\alpha A_{k\alpha} N_{\alpha j} \log(z_\alpha - c_\alpha) + \sum_\alpha \bar{A}_{k\alpha} \bar{N}_{\alpha j} \log(\bar{z}_\alpha - \bar{c}_\alpha) \right\} d_j$$

$$\quad - \frac{1}{4\pi} \left\{ \sum_\alpha A_{k\alpha} N_{\alpha q} \sum_\beta \bar{A}_{q\beta} \bar{N}_{\beta j} \log(z_\alpha - \bar{c}_\beta) \right.$$

$$\quad \left. + \sum_\alpha \bar{A}_{k\alpha} \bar{N}_{\alpha q} \sum_\beta A_{q\beta} N_{\beta j} \log(\bar{z}_\alpha - c_\beta) \right\} d_j. \tag{5.3.6}$$

Use of (1.2.12) in (5.3.6) shows that the boundary condition $\phi_k(x_1, 0)$ is satisfied. Thus (5.3.6) is the required solution to (5.1.1).

5.4   Second problem for a half-plane

A solution to (5.1.1) is required which is valid in the half-plane $x_2 < 0$ and which satisfies the boundary condition

$$P_k(x_1, 0) = 0. \tag{5.4.1}$$

Now on the boundary of the half-space $x_2 < 0$, $n_1 = 0$ and $n_2 = 1$ so (1.2.6) yields

$$P_k(x_1, 0) = \psi_{k2}(x_1, 0). \tag{5.4.2}$$

To construct the required solution consider again the functions (5.3.2) with the corresponding ϕ_k given by (5.3.3). The d_j in (5.3.3) are required to be related to the K_i by (5.3.4) so that equation (5.3.3) yields a solution to (5.1.1). Substitution of (5.3.2) into (1.2.7) yields

$$\psi_{ij} = \frac{1}{4\pi} \left\{ \sum_\alpha \frac{L_{ij\alpha} N_{\alpha k}}{z_\alpha - c_\alpha} + \sum_\alpha \frac{\bar{L}_{ij\alpha} \bar{N}_{\alpha k}}{\bar{z}_\alpha - \bar{c}_\alpha} \right\} d_k$$

$$\quad + \sum_\alpha L_{ij\alpha} g'_\alpha(z_\alpha) + \sum_\alpha \bar{L}_{ij\alpha} \bar{g}'_\alpha(\bar{z}_\alpha). \tag{5.4.3}$$

The $g_\alpha(z_\alpha)$ in (5.3.4) are analytic in $x_2 < 0$ and must be chosen so that the boundary condition (5.4.2) is satisfied. Image considerations lead to the choice

$$g_\alpha(z_\alpha) = -\frac{1}{4\pi} \left\{ \sum_\alpha M_{\alpha k} \sum_\beta \bar{L}_{k2\beta} \bar{N}_{\beta j} \log(z_\alpha - \bar{c}_\beta) \right\} d_j. \tag{5.4.4}$$

Substitution of (5.4.4) into (5.4.3) yields

$$\psi_{ij} = \frac{1}{4\pi} \left\{ \sum_\alpha \frac{L_{ij\alpha} N_{\alpha k}}{z_\alpha - c_\alpha} + \sum_\alpha \frac{\bar{L}_{ij\alpha} \bar{N}_{\alpha k}}{\bar{z}_\alpha - \bar{c}_\alpha} \right\} d_k$$

$$- \frac{1}{4\pi} \left\{ \sum_\alpha L_{ij\alpha} M_{\alpha k} \sum_\alpha \bar{L}_{k2\beta} \bar{N}_{\beta r} (z_\alpha - \bar{c}_\beta)^{-1} \right.$$

$$\left. + \sum_\alpha \bar{L}_{ij\alpha} \bar{M}_{\alpha k} \sum_\beta L_{k2\beta} N_{\beta r} (\bar{z}_\alpha - c_\beta)^{-1} \right\} d_r. \tag{5.4.5}$$

Use of (1.2.20) shows that the boundary condition $P_i(x_1, 0) = \psi_{i2}(x_1, 0) = 0$ is satisfied by (5.4.5).

5.5 Mixed boundary-value problems for a half-plane

Two mixed boundary-value problems for the half-plane $x_2 < 0$ will be considered in this section. In both cases a solution to (5.1.1) is required which is valid in $x_2 < 0$. The boundary conditions on the boundary $x_2 = 0$ of the half-plane are as follows.

Problem I

$$\phi_i(x_1, 0) = 0 \quad \text{for} \quad x_1 < a \quad \text{and} \quad x_1 > b, \tag{5.5.1}$$

$$P_i(x_1, 0) = \psi_{i2}(x_1, 0) = 0 \quad \text{for} \quad a < x_1 < b. \tag{5.5.2}$$

Problem II

$$P_i(x_1, 0) = \psi_{i2}(x_1, 0) = 0 \quad \text{for} \quad x_1 < a \quad \text{and} \quad x_1 > b, \tag{5.5.3}$$

$$\phi_i(x_1, 0) = 0 \quad \text{for} \quad a < x_1 < b. \tag{5.5.4}$$

To find the required solution for Problem I consider the expression (5.3.6) for ϕ_k. That is

$$\phi_k^{(1)} = \frac{1}{4\pi} \left\{ \sum_\alpha A_{k\alpha} N_{\alpha j} \log(z_\alpha - c_\alpha) + \sum_\alpha \bar{A}_{k\alpha} \bar{N}_{\alpha j} \log(\bar{z}_\alpha - \bar{c}_\alpha) \right\} d_j$$

$$- \frac{1}{4\pi} \left\{ \sum_\alpha A_{k\alpha} N_{\alpha q} \sum_\beta \bar{A}_{q\beta} \bar{N}_{\beta j} \log(z_\alpha - \bar{c}_\beta) \right.$$

$$\left. + \sum_\alpha \bar{A}_{k\alpha} \bar{N}_{\alpha q} \sum_\beta A_{q\beta} N_{\beta j} \log(\bar{z}_\alpha - c_\beta) \right\} d_j. \tag{5.5.5}$$

If the constants d_i are determined by (5.2.6) then this is a solution to (5.1.1) which certainly satisfies condition (5.5.1). However, on $x_2 = 0$ the solution yields

$$\psi_{i2}^{(1)}(x, 0) = \frac{1}{4\pi} \left\{ \sum_\alpha \frac{L_{i2\alpha}N_{\alpha k}}{x_1 - c_\alpha} + \sum_\alpha \frac{\bar{L}_{i2\alpha}\bar{N}_{\alpha k}}{x_1 - \bar{c}_\alpha} \right\} d_k$$

$$- \frac{1}{4\pi} \left\{ \sum_\alpha L_{i2\alpha}N_{\alpha k} \sum_\beta \bar{A}_{k\beta}\bar{N}_{\beta r}(x_1 - \bar{c}_\beta)^{-1} \right.$$

$$\left. + \sum_\alpha \bar{L}_{i2\alpha}\bar{N}_{\alpha k} \sum_\beta A_{k\beta}N_{\beta r}(x_1 - c_\beta)^{-1} \right\} d_r. \qquad (5.5.6)$$

$$= \frac{1}{4\pi} \sum_\alpha \left\{ \frac{L_{i2\alpha}N_{\alpha k} - \bar{C}_{ir}A_{r\alpha}N_{\alpha k}}{x_1 - c_\alpha} \right\} d_k$$

$$+ \frac{1}{4\pi} \sum_\alpha \left\{ \frac{\bar{L}_{i2\alpha}\bar{N}_{\alpha k} - C_{ir}\bar{A}_{r\alpha}\bar{N}_{\alpha k}}{x_1 - \bar{c}_\alpha} \right\} d_k$$

$$= \sum_\alpha \left\{ \frac{\mathscr{W}_{i\alpha}}{x_1 - c_\alpha} + \frac{\bar{\mathscr{W}}_{i\alpha}}{x_1 - \bar{c}_\alpha} \right\}, \qquad (5.5.7)$$

where

$$\mathscr{W}_{i\alpha} = \frac{1}{4\pi} \left\{ L_{i2\alpha}N_{\alpha k} - \bar{C}_{ir}A_{r\alpha}N_{\alpha k} \right\} d_k \qquad (5.5.8)$$

and C_{ir} is defined by (1.2.17).

Now the required solution is such that $\psi_{i2}(x_1, 0) = 0$ for $a < x_1 < b$ and in order to achieve this condition a suitable solution of (1.1.1) is added to (5.5.5). This function may be obtained as follows. In Section 3.4 put $g_k(x_1) = 0$ and $p_i(x_1) = -\psi_{i2}^{(1)}(x_1, 0)$ where $\psi_{i2}^{(1)}(x_1, 0)$ is given by (5.5.6). Thus from (3.4.15)

$$\Psi_\gamma(z) = -\frac{R_{\gamma i}}{2\pi i} \int_a^b \frac{p_i(t)\, dt}{X_\gamma^+(t)(t - z)}$$

$$= \frac{R_{\gamma i}}{(1 - e^{-2\pi i m})} \sum_\alpha \left\{ \frac{\mathscr{W}_{i\alpha}}{z - c_\alpha} [X_\gamma^{-1}(z)(1 - X_\gamma(z)\{(m-1)b + ma + z\}) \right.$$

$$- X_\gamma^{-1}(c_\alpha)(1 - X_\gamma(c_\alpha)\{(m-1)b + ma + c_\alpha\})]$$

$$+ \frac{\bar{\mathscr{W}}_{i\alpha}}{z - \bar{c}_\alpha} [X_\gamma^{-1}(z)(1 - X_\gamma(z)\{(m-1)b + ma + z\})$$

$$\left. - X_\gamma^{-1}(\bar{c}_\alpha)(1 - X_\gamma(\bar{c}_\alpha)\{(m-1)b + ma + \bar{c}_\alpha\})] \right\}, \qquad (5.5.9)$$

where

$$X_\gamma(z) = (z - b)^{m-1}(z - a)^{-m} \qquad (5.5.10)$$

with

$$m = \frac{1}{2\pi i} \log \lambda_\gamma. \tag{5.5.11}$$

The λ_γ in (5.5.11) are the roots of the equation (3.4.9). Now from (3.4.11) and (3.4.18)

$$\theta_j'(z) = \sum_\gamma T_{j\gamma} X_\gamma(z) \Psi_\gamma(z). \tag{5.5.12}$$

The required functions ϕ_k and ψ_{ij} which must be added to (5.5.5) and (5.5.6) so that the boundary condition (5.5.2) is satisfied are given by (3.2.4) and (3.2.5). Specifically

$$\phi_k^{(2)} = \sum_\alpha A_{k\alpha} N_{\alpha j} \theta_j(z_\alpha) - \sum_\alpha \bar{A}_{k\alpha} \bar{N}_{\alpha j} \theta_j(\bar{z}_\alpha) \quad \text{for} \quad x_2 < 0, \tag{5.5.13}$$

$$\psi_{ij}^{(2)} = \sum_\alpha L_{ij\alpha} N_{\alpha k} \theta_k'(z_\alpha) - \sum_\alpha \bar{L}_{ij\alpha} \bar{N}_{\alpha k} \theta_k'(\bar{z}_\alpha) \quad \text{for} \quad x_2 < 0. \tag{5.5.14}$$

The required solution to Problem I is thus given by

$$\phi_k = \phi_k^{(1)} + \phi_k^{(2)} \tag{5.5.15}$$

$$\psi_{ij} = \psi_{ij}^{(1)} + \psi_{ij}^{(2)}. \tag{5.5.16}$$

These expressions for ϕ_k and ψ_{ij} are rather complicated so it may be instructive to examine their form for the simple case of Laplace's equation. In this case all the constants occurring in the preceding equations are zero except for those listed below (together with their conjugates)

$$\tau_1 = i, \quad A_{11} = 1, \quad L_{111} = K, \quad L_{121} = iK, \quad N_{11} = 1,$$

$$M_{11} = -iK^{-1}, \quad C_{11} = iK, \quad B_{11} = -iK^{-1}, \quad \lambda = -1,$$

$$m = \tfrac{1}{2}, \quad R_{11} = 1, \quad S_{11} = -iK, \quad T_{11} = iK^{-1}, \tag{5.5.17}$$

where K is an arbitrary constant (see Section 2.2). Hence, (5.5.5) and (5.5.6) yield

$$\phi_1^{(1)} = \frac{1}{4\pi} \{\log (z_1 - c_1) + \log (\bar{z}_1 - \bar{c}_1) - \log (z_1 - \bar{c}_1)$$

$$- \log (\bar{z}_1 - c_1)\} d_1, \tag{5.5.18}$$

$$\psi_{12}^{(1)}(x_1, 0) = \frac{iK}{2\pi} \left\{ \frac{1}{x_1 - c_1} - \frac{1}{x_1 - \bar{c}_1} \right\} d_1. \tag{5.5.19}$$

Also (5.5.9)–(5.5.11) yield

$$\Psi_1(z) = \frac{1}{2} \left\{ \frac{\mathcal{W}_{11}}{z - c_1} \left[\frac{1 - X(z)(\tfrac{1}{2}(a+b) + z)}{X(z)} - \frac{1 - X(c_1)(\tfrac{1}{2}(a+b) + c_1)}{X(c_1)} \right] \right.$$

$$\left. + \frac{\bar{\mathcal{W}}_{11}}{z - \bar{c}_1} \left[\frac{1 - X(z)(\tfrac{1}{2}(a+b) + z)}{X(z)} - \frac{1 - X(\bar{c}_1)(\tfrac{1}{2}(a+b) + \bar{c}_1)}{X(\bar{c}_1)} \right] \right\}, \tag{5.5.20}$$

$$X(z) = (z - b)^{-1/2}(z - a)^{-1/2}. \tag{5.5.21}$$

For simplicity, let $a = -1$ and $b = 1$. Then, upon integration, (5.5.12) yields

$$
\theta_1(z) = \frac{i}{2K}\left\{ \mathcal{W}_{11}\left[\log(z - c_1) - \cosh^{-1} z - \frac{c_1}{\sqrt{c_1^2 - 1}} \right.\right.
$$

$$
\times \log\left[\frac{z + \sqrt{z^2 - 1} - c_1 - \sqrt{c_1^2 - 1}}{z + \sqrt{z^2 - 1} - c_1 + \sqrt{c_1^2 - 1}} \right]
$$

$$
\left. - \frac{1}{\sqrt{c_1^2 - 1}} \log\left[\frac{z + \sqrt{z^2 - 1} - c_1 - \sqrt{c_1^2 - 1}}{z + \sqrt{z^2 - 1} - c_1 + \sqrt{c_1^2 - 1}} \right]\left(\frac{1 - c_1 X(c_1)}{X(c_1)} \right) \right]
$$

$$
+ \bar{\mathcal{W}}_{11}\left[\log(z - \bar{c}_1) - \cosh^{-1} z - \frac{\bar{c}_1}{\sqrt{\bar{c}_1^2 - 1}} \right.
$$

$$
\times \log\left[\frac{z + \sqrt{z^2 - 1} - \bar{c}_1 - \sqrt{\bar{c}_1^2 - 1}}{z + \sqrt{z^2 - 1} - \bar{c}_1 + \sqrt{\bar{c}_1^2 - 1}} \right]
$$

$$
\left.\left. - \frac{1}{\sqrt{\bar{c}^2 - 1}} \log\left[\frac{z + \sqrt{z^2 - 1} - \bar{c}_1 - \sqrt{\bar{c}_1^2 - 1}}{z + \sqrt{z^2 - 1} - \bar{c}_1 + \sqrt{\bar{c}_1^2 - 1}} \right]\left(\frac{1 - \bar{c}_1 X(\bar{c}_1)}{X(\bar{c}_1)} \right) \right] \right\}.
$$

$$(5.5.22)$$

Also

$$
\theta_1'(z) = \frac{i}{2K}\left\{ \frac{\mathcal{W}_{11}}{z - c_1}\left[1 - z X_1(z) - X_1(z)\left(\frac{1 - c_1 X_1(c_1)}{X_1(c_1)} \right) \right] \right.
$$

$$
\left. + \frac{\bar{\mathcal{W}}_{11}}{z - \bar{c}_1}\left[1 - z X_1(z) - X_1(z)\left(\frac{1 - \bar{c}_1 X_1(\bar{c}_1)}{X_1(\bar{c}_1)} \right) \right] \right\}.
$$

$$(5.5.23)$$

In the above equations $\mathcal{W}_{11}$ is given by

$$
\mathcal{W}_{11} = \frac{iKd_1}{2\pi}.
$$

$$(5.5.24)$$

From (5.5.13) and (5.5.14) it follows that

$$
\phi_1^{(2)} = \theta_1(z) - \theta_1(\bar{z}),
$$

$$(5.5.25)$$

$$
\psi_{12}^{(2)} = iK[\theta_1'(z) + \theta_1'(\bar{z})].
$$

$$(5.5.26)$$

From (5.5.23) and (5.5.26) it is apparent that

$$
\psi_{12}^{(2)}(x_1, 0) = -\frac{iK}{2\pi}\left\{ \frac{1}{x_1 - c_1} - \frac{1}{x_1 - \bar{c}_1} \right\} \quad \text{for} \quad |x_1| < 1.
$$

Hence

$$\psi_{12}(x_1, 0) = \psi_{12}^{(1)}(x_1, 0) + \psi_{12}^{(2)}(x_1, 0)$$
$$= 0 \quad \text{for} \quad |x_1| < 1$$

thus showing that condition (5.5.2) is satisfied. Similarly, from (5.5.18), (5.5.22) and (5.5.25) it is apparent that condition (5.5.1) is satisfied.

Now consider Problem II. To find the required solution to this problem consider the expression (5.3.3) for ϕ_k with $g_\alpha(z)$ given by (5.4.4). That is

$$\phi_k^{(1)} = \frac{1}{4\pi} \left\{ \sum_\alpha A_{k\alpha} N_{\alpha j} \log (z_\alpha - c_\alpha) + \sum_\alpha \bar{A}_{k\alpha} \bar{N}_{\alpha j} \log (\bar{z}_\alpha - \bar{c}_\alpha) \right\} d_j$$

$$- \frac{1}{4\pi} \left\{ \sum_\alpha A_{k\alpha} M_{\alpha q} \sum_\beta \bar{L}_{q2\beta} \bar{N}_{\beta j} \log (z_\alpha - \bar{c}_\beta) \right.$$

$$\left. + \sum_\alpha \bar{A}_{k\alpha} \bar{M}_{\alpha q} \sum_\beta L_{q2\beta} N_{\beta j} \log (\bar{z}_\alpha - c_\beta) \right\} d_j, \tag{5.5.27}$$

$$\psi_{ij}^{(1)} = \frac{1}{4\pi} \left\{ \sum_\alpha \frac{L_{ij\alpha} N_{\alpha k}}{z_\alpha - c_\alpha} - \sum_\alpha \frac{\bar{L}_{ij\alpha} \bar{N}_{\alpha k}}{\bar{z}_\alpha - \bar{c}_\alpha} \right\} d_k$$

$$- \frac{1}{4\pi} \left\{ \sum_\alpha L_{ij\alpha} M_{\alpha k} \sum_\beta \bar{L}_{k2\beta} \bar{N}_{\beta r} (z_\alpha - \bar{c}_\beta)^{-1} \right.$$

$$\left. + \sum_\alpha \bar{L}_{ij\alpha} \bar{M}_{\alpha k} \sum_\beta L_{k2\beta} N_{\beta r} (\bar{z}_\alpha - c_\beta)^{-1} \right\} d_r. \tag{5.5.28}$$

This solution to (5.1.1) satisfies the condition (5.5.3). On $x_2 = 0$ (5.5.27) yields

$$\phi_k^{(1)\prime}(x_1, 0) = \frac{1}{4\pi} \left\{ \sum_\alpha \frac{A_{k\alpha} N_{\alpha j}}{x_1 - c_\alpha} + \sum_\alpha \frac{\bar{A}_{k\alpha} \bar{N}_{\alpha j}}{x_1 - \bar{c}_\alpha} \right\} d_j$$

$$- \frac{1}{4\pi} \left\{ \sum_\alpha A_{k\alpha} M_{\alpha q} \sum_\beta \bar{L}_{q2\beta} \bar{N}_{\beta r} (x_1 - \bar{c}_\beta)^{-1} \right.$$

$$\left. + \sum_\alpha \bar{A}_{k\alpha} \bar{M}_{\alpha q} \sum_\beta L_{q2\beta} N_{\beta j} (x_1 - c_\beta)^{-1} \right\} d_j$$

$$= \frac{1}{4\pi} \sum_\alpha \left\{ \frac{A_{k\alpha} N_{\alpha j} - \bar{B}_{kq} L_{q2\alpha} N_{\alpha j}}{x_1 - c_\alpha} \right\} d_j$$

$$+ \frac{1}{4\pi} \sum_\alpha \left\{ \frac{\bar{A}_{k\alpha} \bar{N}_{\alpha j} - B_{kq} \bar{L}_{q2\alpha} \bar{N}_{\alpha r}}{x_1 - \bar{c}_\alpha} \right\} d_j$$

$$= \sum_\alpha \left\{ \frac{\mathcal{V}_{k\alpha}}{x_1 - c_\alpha} + \frac{\bar{\mathcal{V}}_{k\alpha}}{x_1 - \bar{c}_\alpha} \right\}, \tag{5.5.29}$$

where

$$\mathcal{V}_{k\alpha} = \frac{1}{4\pi} \{A_{k\alpha}N_{\alpha j} - \bar{B}_{kq}L_{q2\alpha}N_{\alpha j}\}d_j \tag{5.5.30}$$

and B_{kq} is defined by (1.2.25).

The required solution to (5.1.1) is such that $\phi_k(x_1, 0) = 0$ for $a < x_1 < b$. To satisfy this condition it is advantageous to proceed as in Problem I and add a suitable solution of (1.1.1) to (5.5.27). The required function may be obtained from Section 3.5 by putting $p_i(x_1) = 0$ and $g'_k(x_1) = -\phi^{(1)\prime}(x_1, 0)$ given by (5.5.29). Then from (3.5.4)

$$\chi'_r(z) = \frac{\sum_{\alpha} T_{r\gamma}R_{\gamma k}X_{\gamma}(z)}{2\pi i} \int_a^b \frac{\phi_k^{(1)\prime}(t, 0)\,dt}{X_{\gamma}^+(t)(t-z)}$$

$$= \frac{\sum_{\alpha} T_{r\gamma}R_{\gamma k}X_{\gamma}(z)}{(1-e^{-2\pi i m})} \sum_{\alpha} \left\{ \frac{\mathcal{V}_{k\alpha}}{z-c_{\alpha}} [X_{\gamma}^{-1}(z) \right.$$

$$\times (1 - X_{\gamma}(z)\{(m-1)b + ma + z\})$$

$$- X_{\gamma}^{-1}(c_{\alpha})(1 - X_{\gamma}(c_{\alpha})\{(m-1)b + ma + c_{\alpha}\})]$$

$$+ \frac{\mathcal{V}_{k\alpha}}{z-\bar{c}_{\alpha}} [X_{\gamma}^{-1}(z)(1 - X_{\gamma}(z)\{(m-1)b + ma + z\})$$

$$\left. - X_{\gamma}^{-1}(\bar{c}_{\alpha})(1 - X_{\gamma}(\bar{c}_{\alpha})\{(m-1)b + ma + \bar{c}_{\alpha}\})] \right\}, \tag{5.5.31}$$

where $R_{\gamma k}$, $T_{r\gamma}$, $X_{\gamma}(z)$ and m are defined in Section 3.5. Thus, the function which must be added to (5.5.27) so that the boundary condition (5.5.4) is satisfied is

$$\phi_k^{(2)} = \sum_{\alpha} A_{k\alpha}M_{\alpha j}\chi_j(z_{\alpha}) - \sum_{\alpha} \bar{A}_{k\alpha}\bar{M}_{\alpha j}\chi_j(\bar{z}_{\alpha}) \quad \text{for} \quad x_2 < 0, \tag{5.5.32}$$

where $\chi_j(z)$ is obtained by integrating (5.5.31). The corresponding expression for ψ_{ij} is

$$\psi_{ij}^{(2)} = \sum_{\alpha} L_{ij\alpha}M_{\alpha k}\chi'_k(z_{\alpha}) - \sum_{\alpha} \bar{L}_{ij\alpha}\bar{M}_{\alpha k}\chi'_k(\bar{z}_{\alpha}) \quad \text{for} \quad x_2 < 0. \tag{5.5.33}$$

The required solution to Problem II is now given by

$$\phi_k = \phi_k^{(1)} + \phi_k^{(2)}, \tag{5.5.34}$$

$$\psi_{ij} = \psi_{ij}^{(1)} + \psi_{ij}^{(2)}. \tag{5.5.35}$$

5.6 Problem for the cut plane

Consider the whole plane cut along the x_1-axis from $x_1 = -1$ to $x_1 = 1$. A solution to (5.1.1) is required which is valid in the cut plane and satisfies the boundary condition

$$P_k(x_1, 0) = 0 \quad \text{for} \quad |x_1| < 1. \tag{5.6.1}$$

To find the required solution to this problem consider the expression (5.2.5) for ϕ_k. That is

$$\phi_k^{(1)} = \frac{1}{4\pi} \left\{ \sum_\alpha A_{k\alpha} N_{\alpha j} \log(z_\alpha - c_\alpha) + \sum_\alpha \bar{A}_{k\alpha} \bar{N}_{\alpha j} \log(\bar{z}_\alpha - \bar{c}_\alpha) \right\} d_j, \tag{5.6.2}$$

where the constants d_j are given by (5.2.6). The corresponding ψ_{ij} is

$$\psi_{ij}^{(1)} = \frac{1}{4\pi} \left\{ \sum_\alpha L_{ij\alpha} N_{\alpha k} (z_\alpha - c_\alpha)^{-1} + \sum_\alpha \bar{L}_{ij\alpha} \bar{N}_{\alpha k} (\bar{z}_\alpha - \bar{c}_\alpha)^{-1} \right\} d_k. \tag{5.6.3}$$

On $x_2 = 0$

$$\psi_{i2}^{(1)} = \frac{1}{4\pi} \left\{ \sum_\alpha L_{i2\alpha} N_{\alpha k} (x_1 - c_\alpha)^{-1} + \sum_\alpha \bar{L}_{i2\alpha} \bar{N}_{\alpha k} (x_1 - \bar{c}_\alpha)^{-1} \right\} d_k. \tag{5.6.4}$$

Hence, since $P_i = \psi_{ij} n_j$, it is clear that the boundary condition (5.6.1) is not satisfied by the solution (5.6.2). In order to negate this ψ_{i2} and hence satisfy the condition (5.6.1) a new solution $\phi_k^{(2)}$ is added to (5.6.2). Now if

$$\psi_{i2} = p_i(x_1) \tag{5.6.5}$$

over the crack faces then, from equations (3.7.10), (3.7.24) and (3.7.26),

$$\Psi_i'(z) = \frac{X(z)}{2\pi i} \int_{-1}^1 \frac{p_i(t)\, dt}{X^+(t)(t - z)} \tag{5.6.6}$$

where

$$X(z) = (z^2 - 1)^{-1/2}. \tag{5.6.7}$$

From (5.6.4) the function $p_i(x_1)$ is given by

$$p_i(x_1) = -\psi_{i2}^{(1)}(x_1, 0)$$

$$= -\frac{1}{4\pi} \left\{ \sum_\alpha L_{i2\alpha} N_{\alpha k} (x_1 - c_\alpha)^{-1} + \sum_\alpha \bar{L}_{i2\alpha} \bar{N}_{\alpha k} (x_1 - \bar{c}_\alpha)^{-1} \right\} d_k. \tag{5.6.8}$$

Substitution of (5.6.8) into (5.6.6) yields

$$\Psi_i'(z) = -\frac{X(z)}{8\pi^2} \left\{ \sum_\alpha L_{i2\alpha} N_{\alpha k} \int_{-1}^1 \frac{(1 - t^2)^{1/2}\, dt}{(t - c_\alpha)(t - z)} \right.$$

$$\left. + \sum_\alpha \bar{L}_{i2\alpha} \bar{N}_{\alpha k} \int_{-1}^1 \frac{(1 - t^2)^{1/2}\, dt}{(t - \bar{c}_\alpha)(t - z)} \right\} d_k. \tag{5.6.9}$$

Partial fractions and contour integration may be used to show that

$$\int_{-1}^{1} \frac{(1-t^2)^{1/2}\,dt}{(t-k)(t-z)} = -\frac{\pi}{z-k}\{z-k+(k^2-1)^{1/2}-(z^2-1)^{1/2}\}, \qquad (5.6.10)$$

where k is a constant. Use of this result in (5.6.9) yields

$$\Psi_i'(z) = \frac{X(z)}{8\pi}\left\{\sum_\alpha L_{i2\alpha}N_{\alpha k}\left[1+\frac{(c_\alpha^2-1)^{1/2}-(z^2-1)^{1/2}}{z-c_\alpha}\right]\right.$$
$$\left. +\sum_\alpha \bar{L}_{i2\alpha}\bar{N}_{\alpha k}\left[1+\frac{(\bar{c}_\alpha^2-1)^{1/2}-(z^2-1)^{1/2}}{z-\bar{c}_\alpha}\right]\right\}d_k. \qquad (5.6.11)$$

Hence

$$\Psi_i(z) = -\frac{1}{8\pi}\left\{\sum_\alpha L_{i2\alpha}N_{\alpha k}\Omega(z,c_\alpha)+\sum_\alpha \bar{L}_{i2\alpha}\bar{N}_{\alpha k}\Omega(z,\bar{c}_\alpha)\right\}d_k \qquad (5.6.12)$$

where

$$\Omega(z,c_\alpha) = \log(z-c_\alpha)-\log\left[\frac{z+(z^2-1)^{1/2}-c_\alpha-(c_\alpha^2-1)^{1/2}}{z+(z^2-1)^{1/2}-c_\alpha+(c_\alpha^2-1)^{1/2}}\right]-\cosh^{-1}z.$$
$$(5.6.13)$$

Equations (3.7.5)–(3.7.8) now yield

$$\phi_k^{(2)} = -\frac{1}{4\pi}\,\mathcal{R}\sum_\alpha A_{k\alpha}M_{\alpha j}\left\{\sum_\beta L_{j2\beta}N_{\beta k}\Omega(z_\alpha,c_\beta)\right.$$
$$\left. +\sum_\beta \bar{L}_{j2\beta}\bar{N}_{\beta k}\Omega(z_\alpha,\bar{c}_\beta)\right\}d_k, \qquad (5.6.14)$$

$$\psi_{ij}^{(2)} = -\frac{1}{4\pi}\,\mathcal{R}\sum_\alpha L_{ij\alpha}M_{\alpha k}(z_\alpha^2-1)^{-\frac{1}{2}}\left\{\sum_\beta \{L_{k2\beta}N_{\beta r}\chi(z_\alpha,c_\beta)\right.$$
$$\left. +\bar{L}_{k2\beta}\bar{N}_{\beta r}\chi(z_\alpha,\bar{c}_\beta)\}d_r\right\}, \qquad (5.6.15)$$

where

$$\chi(z_\alpha,c_\beta) = \frac{(z_\alpha^2-1)^{1/2}-(c_\beta^2-1)^{1/2}}{z_\alpha-c_\beta}-1. \qquad (5.6.16)$$

5.7 Problem for a strip between two parallel lines

Consider the strip in the x_1x_2-plane lying between the lines $x_2=\pm h$. A solution to (5.1.1) is required which is valid in the strip and satisfies the boundary condition

$$P_i(x_1,\pm h) = 0. \qquad (5.7.1)$$

To construct the solution to this problem consider the solution (5.2.3) to (5.1.1). That is

$$\phi_k^{(1)} = \frac{1}{4\pi} \left\{ \sum_\alpha A_{k\alpha} N_{\alpha j} \log (z_\alpha - c_\alpha) + \sum_\alpha \bar{A}_{k\alpha} \bar{N}_{\alpha j} \log (\bar{z}_\alpha - \bar{c}_\alpha) \right\} d_j \quad (5.7.2)$$

where the d_j are given by (5.2.6). also

$$\psi_{ij}^{(1)} = \frac{1}{4\pi} \sum_\alpha \left\{ \frac{L_{ij\alpha} N_{\alpha k}}{z_\alpha - c_\alpha} + \frac{\bar{L}_{ij\alpha} \bar{N}_{\alpha k}}{\bar{z}_\alpha - \bar{c}_\alpha} \right\} d_k. \quad (5.7.3)$$

On $x_2 = \pm h$ this equation yields

$$\psi_{i2}^{(1)}(x_1, h) = \frac{1}{4\pi} \sum_\alpha \left\{ \frac{L_{i2\alpha} N_{\alpha k}}{x_1 + \tau_\alpha h - c_\alpha} + \frac{\bar{L}_{i2\alpha} \bar{N}_{\alpha k}}{x_1 + \bar{\tau}_\alpha h - \bar{c}_\alpha} \right\} d_k, \quad (5.7.4)$$

$$\psi_{i2}^{(1)}(x_1, -h) = \frac{1}{4\pi} \sum_\alpha \left\{ \frac{L_{i2\alpha} N_{\alpha k}}{x_1 - \tau_\alpha h - c_\alpha} + \frac{\bar{L}_{i2\alpha} \bar{N}_{\alpha k}}{x_1 - \bar{\tau}_\alpha h - \bar{c}_\alpha} \right\} d_k. \quad (5.7.5)$$

Now the boundary condition (5.7.1) will be satisfied if $\psi_{i2}(x_1, \pm h) = 0$. In order to satisfy this condition it is necessary to add a solution of (1.1.1) to (5.7.2). This solution is used to negate the contribution of (5.7.3) to the value of $\psi_{i2}^{(1)}$ on the boundaries $x_2 = \pm h$ of the strip.

From Section 3.9 it follows that the required solution to (1.1.1) is given by

$$\phi_k^{(2)} = \sum_\alpha A_{k\alpha} f_\alpha(z_\alpha) + \sum_\alpha \bar{A}_{k\alpha} \bar{f}_\alpha(\bar{z}_\alpha), \quad (5.7.6)$$

$$\psi_{ij}^{(2)} = \sum_\alpha L_{ij\alpha} f'_\alpha(z_\alpha) + \sum_\alpha \bar{L}_{ij\alpha} \bar{f}'_\alpha(\bar{z}_\alpha), \quad (5.7.7)$$

where

$$f_\alpha(z_\alpha) = \frac{1}{2\pi} \int_0^\infty \{E_\alpha(p) \exp (ipz_\alpha) + F_\alpha(p) \exp (-ipz_\alpha)\} \, dp. \quad (5.7.8)$$

The functions $E_\alpha(p)$ and $F_\alpha(p)$ in (5.7.8) are given (from (3.9.9) and (3.9.11)) by

$$\sum_\alpha [S_{i\alpha} E_\alpha + \bar{R}_{i\alpha} \bar{F}_\alpha] = \frac{-i}{p} \int_{-\infty}^\infty \psi_{i2}^{(1)}(\xi, h) \exp (-ip\xi) \, d\xi, \quad (5.7.9)$$

$$\sum_\alpha [R_{i\alpha} E_\alpha + \bar{S}_{i\alpha} \bar{F}_\alpha] = \frac{-i}{p} \int_{-\infty}^\infty \psi_{i2}^{(1)}(\xi, -h) \exp (-ip\xi) \, d\xi. \quad (5.7.10)$$

Substitution of (5.7.4) and (5.7.5) into the integrands and integration of the resulting expressions yields

$$\sum_\alpha [S_{i\alpha} E_\alpha + \bar{R}_{i\alpha} \bar{F}_\alpha] = (2p)^{-1} d_k \sum_\alpha L_{i2\alpha} N_{\alpha k} \exp [ip(\tau_\alpha h - c_\alpha)], \quad (5.7.11)$$

$$\sum_{\alpha} [R_{i\alpha} E_{\alpha} + \bar{S}_{i\alpha} \bar{F}_{\alpha}] = (2p)^{-1} d_k \sum_{\alpha} \bar{L}_{i2\alpha} \bar{N}_{\alpha k} \exp\left[-ip(\bar{\tau}_{\alpha} h + \bar{c}_{\alpha})\right]. \quad (5.7.12)$$

Equations (5.7.11) and (5.7.12) constitute $2N$ simultaneous linear equations for the $2N$ unknowns E_{α}, $\bar{F}_{\alpha}$, $\alpha = 1, 2, \ldots, N$. These equations may be solved to find E_{α} and F_{α} and then (5.7.6)–(5.7.8) yield the $\phi_k^{(2)}$ and $\psi_{ij}^{(2)}$. The required solution to (5.1.1) is then given by

$$\phi_k = \phi_k^{(1)} + \phi_k^{(2)},$$
$$\psi_{ij} = \psi_{ij}^{(1)} + \psi_{ij}^{(2)}.$$

6

Boundary integral equations

6.1 Introduction

Consider the inhomogeneous system obtained by replacing the right hand side of (1.1.1) by known functions $h_i(x_1, x_2)$. The system is

$$a_{ijkl} \frac{\partial^2 \phi_k}{\partial x_j \, \partial x_l} = h_i(x_1, x_2) \quad \text{for} \quad i = 1, 2, \ldots, N. \tag{6.1.1}$$

In the present chapter a reciprocal theorem is introduced for the purpose of relating any two solutions of the system (6.1.1). Now the homogeneous system (1.1.1) and the inhomogeneous system (5.1.1) are both special cases of (6.1.1). Hence the reciprocal theorem may be used to relate a particular solution of (5.1.1) to any desired solution of (1.1.1). Specifically, the reciprocal theorem relates a particular solution of (5.1.1) to the solution of a boundary-value problem governed by (1.1.1) through an integral taken over the boundary of the region under consideration. Further, the boundary integral can often be considerably simplified if the particular solution of (5.1.1) is suitably chosen. Thus, after deriving the reciprocal theorem various boundary integral equations are derived. The first is for the general boundary-value problem specified in Section 1.1 and involves the fundamental singular solution (5.2.5). The remaining boundary integral formulae are for regions with particular geometries and, in each case, it is possible to obtain a marked simplification in the boundary integral equation by choosing a suitable solution of (5.1.1).

6.2 A reciprocal theorem

The following reciprocal theorem is taken from Clements and Rizzo [12].

Theorem 6.2.1 *Let ϕ_k be a solution to (6.1.1) with corresponding P_k defined by (1.1.4). Let ϕ'_k be a solution to a similar system with h_k replaced by h'_k. Then*

$$\int_C [P_i\phi'_i - P'_i\phi_i]\,\mathrm{d}s = \int_R [h_i\phi'_i - h'_i\phi_i]\,\mathrm{d}R. \qquad (6.2.1)$$

Proof From the definition of P_i and the divergence theorem it follows that

$$\int_C P_i\phi'_i\,\mathrm{d}s = \int_C a_{ijkl}\frac{\partial\phi_k}{\partial x_l}n_j\phi'_i\,\mathrm{d}s$$

$$= \int_R a_{ijkl}\frac{\partial}{\partial x_j}\left(\frac{\partial\phi_k}{\partial x_l}\phi'_i\right)\mathrm{d}R.$$

Use of (6.1.1) yields

$$\int_C P_1\phi'_i\,\mathrm{d}s = \int_R a_{ijkl}\frac{\partial\phi_k}{\partial x_l}\frac{\partial\phi'_i}{\partial x_j}\,\mathrm{d}R + \int_R h_i\phi'_i\,\mathrm{d}R. \qquad (6.2.2)$$

Similarly

$$\int_C P'_i\phi_i\,\mathrm{d}s = \int_C a_{ijkl}\frac{\partial\phi'_k}{\partial x_l}n_j\phi_i\,\mathrm{d}s$$

$$= \int_R a_{ijkl}\frac{\partial}{\partial x_j}\left(\frac{\partial\phi'_k}{\partial x_l}\phi_i\right)\mathrm{d}R$$

$$= \int_R a_{ijkl}\frac{\partial\phi_i}{\partial x_j}\frac{\partial\phi'_k}{\partial x_l}\,\mathrm{d}R + \int_R h'_i\phi_i\,\mathrm{d}R. \qquad (6.2.3)$$

Because of the symmetry properties (1.1.2) the first terms on the right hand sides of (6.2.2) and (6.2.3) are identical. The result (6.2.1) follows immediately.

6.3 The boundary integral equation for the general problem

The general problem posed in Section 1.1 is considered in this section. Specifically, a solution to (1.1.1) is required which is valid in a region R with boundary C. On C either the dependent variables ϕ_k are specified or the P_i are specified. The solution ϕ_k (with corresponding P_k) to this problem is linked to the solution (5.2.5) (denoted by ϕ'_k with corresponding P'_k) of the system (5.1.1) through the reciprocal theorem 6.2.1.

In (5.2.6) let $K_i = \delta_{ij}F$ (where F is an arbitrary constant) for $j = 1, 2, \ldots, N$. The corresponding values of d_r will be denoted by d_{rj}. Then the corresponding dependent variable ϕ_k obtained from (5.2.5) will be denoted by Φ_{kj}. The corresponding P_k obtained from (1.1.4) will be denoted by Γ_{kj}. Thus, in (6.2.1) ϕ_i' is associated with Φ_{ij}, P_i' with Γ_{ij}, h_i' with $K_i \delta(\mathbf{x} - \mathbf{x}_0)$ and the unprimed quantities denote a regular solution to the homogeneous system (1.1.1) (so that $h_i = 0$). Then (6.2.1) yields

$$\int_C [P_i \Phi_{ij} - \Gamma_{ij} \phi_i] \, ds = -F \int_R \phi_j \, \delta(\mathbf{x} - \mathbf{x}_0) \, dR. \qquad (6.3.1)$$

Hence, using the sifting properties of the delta functional (see, for example Stakgold [37])

$$F^{-1} \int_C [P_i(\mathbf{x}) \phi_{ij}(\mathbf{x}, \mathbf{x}_0) - \Gamma_{ij}(\mathbf{x}, \mathbf{x}_0) \phi_i(\mathbf{x})] \, ds(\mathbf{x})$$

$$= - \begin{cases} \phi_j(\mathbf{x}_0) & \text{for} \quad \mathbf{x}_0 \in R, \\ c\phi_j(\mathbf{x}_0) & \text{for} \quad \mathbf{x}_0 \in C, \end{cases} \qquad (6.3.2)$$

where explicit point dependence of the functions is indicated and where the value of c depends on the geometry of the boundary. Specifically if the boundary C has a continuously turning tangent then $c = \frac{1}{2}$. In general, the choice $\phi = 1$ in (6.3.2) yields the equation

$$c = F^{-1} \int_C \Gamma_{ij}(\mathbf{x}, \mathbf{x}_0) \, ds(\mathbf{x}) \quad \text{for} \quad \mathbf{x}_0 \in C. \qquad (6.3.3)$$

For convenience, (6.3.2) may be written as

$$\lambda \phi_j(\mathbf{x}_0) + F^{-1} \int_C [P_i(\mathbf{x}) \Phi_{ij}(\mathbf{x}, \mathbf{x}_0) - \Gamma_{ij}(\mathbf{x}, \mathbf{x}_0) \phi_i(\mathbf{x})] \, ds(\mathbf{x}) = 0, \qquad (6.3.4)$$

where $\lambda = 1$ if $\mathbf{x}_0 \in R$ and $\lambda = c$ if $\mathbf{x}_0 \in C$. Equation (6.3.4) constitutes the boundary integral equation for the general boundary-value problem. If the values of the dependent variable ϕ_i and its derivatives are known on the boundary C of the region under consideration then (6.3.4) provides a solution to the problem in terms of an integral taken over the boundary. The functions Φ_{ij} and Γ_{ij} occurring in this integral are given by

$$\Phi_{ij} = \frac{1}{4\pi} \left\{ \sum_\alpha A_{i\alpha} N_{\alpha r} \log (z_\alpha - c_\alpha) + \sum_\alpha \bar{A}_{i\alpha} \bar{N}_{\alpha r} \log (\bar{z}_\alpha - \bar{c}_\alpha) \right\} d_{rj}, \qquad (6.3.5)$$

$$\Gamma_{ij} = \frac{1}{4\pi} \left\{ \sum_\alpha L_{ir\alpha} N_{\alpha q} (z_\alpha - c_\alpha)^{-1} + \sum_\alpha \bar{L}_{ir\alpha} \bar{N}_{\alpha q} (\bar{z}_\alpha - \bar{c}_\alpha)^{-1} \right\} n_r d_{qi}, \qquad (6.3.6)$$

where

$$\delta_{ij} F = -\tfrac{1}{2} i \sum_\alpha [L_{i2\alpha} N_{\alpha r} - \bar{L}_{i2\alpha} \bar{N}_{\alpha r}] d_{rj}. \qquad (6.3.7)$$

6.4 A particular class of boundary-value problems

Suppose a solution to (1.1.1) is required for a region R with boundary C which may be divided into two parts C_1 and C_2. The part C_1 consists of a section of the line $x_2 = 0$ while the remainder C_2 of the boundary has an arbitrary geometry (Fig. 6.4.1). On C_2 either the dependent variables ϕ_k are specified or the P_i are specified while on C_1 either $\phi_k = 0$ or $P_i = 0$.

Case I $\phi_k = 0$ on C_1. Let the d_{rj} be defined by (6.3.7). The corresponding dependent variable ϕ_k obtained, in this case, from (5.3.6) will be denoted by Φ_{kj} where

$$\begin{aligned}\Phi_{kj} &= \frac{1}{4\pi}\left\{\sum_\alpha A_{k\alpha}N_{\alpha r}\log(z_\alpha - c_\alpha) + \sum_\alpha \bar{A}_{k\alpha}\bar{N}_{\alpha r}\log(\bar{z}_\alpha - \bar{c}_\alpha)\right\}d_{rj} \\ &\quad - \frac{1}{4\pi}\left\{\sum_\alpha A_{k\alpha}N_{\alpha q}\sum_\beta \bar{A}_{q\beta}\bar{N}_{\beta r}\log(z_\alpha - \bar{c}_\beta)\right. \\ &\quad \left. + \sum_\alpha \bar{A}_{k\alpha}\bar{N}_{\alpha q}\sum_\beta A_{q\beta}N_{\beta r}\log(\bar{z}_\alpha - c_\beta)\right\}d_{rj}. \end{aligned} \tag{6.4.1}$$

The corresponding Γ_{ij} is given by

$$\begin{aligned}\Gamma_{ij} &= \frac{1}{4\pi}\left\{\sum_\alpha L_{ir\alpha}N_{\alpha q}(z_\alpha - c_\alpha)^{-1} + \sum_\alpha \bar{L}_{ir\alpha}\bar{N}_{\alpha q}(\bar{z}_\alpha - \bar{c}_\alpha)\right\}n_r d_{qj} \\ &\quad - \frac{1}{4\pi}\left\{\sum_\alpha L_{ir\alpha}N_{\alpha q}\sum_\beta \bar{A}_{q\beta}\bar{N}_{\beta p}(z_\alpha - \bar{c}_\beta)^{-1}\right. \\ &\quad \left. + \sum_\alpha \bar{L}_{ir\alpha}\bar{N}_{\alpha q}\sum_\beta A_{q\beta}N_{\beta p}(\bar{z}_\alpha - c_\beta)^{-1}\right\}n_r\, \mathrm{d}_{pj}. \end{aligned} \tag{6.4.2}$$

In this case (6.2.1) yields

$$\lambda\phi_j(\mathbf{x}_0) + F^{-1}\int_{C=C_1+C_2} [P_i(\mathbf{x})\Phi_{ij}(\mathbf{x}, \mathbf{x}_0) - \Gamma_{ij}(\mathbf{x}, \mathbf{x}_0)\phi_i(\mathbf{x})]\, \mathrm{d}s(\mathbf{x}) = 0, \tag{6.4.3}$$

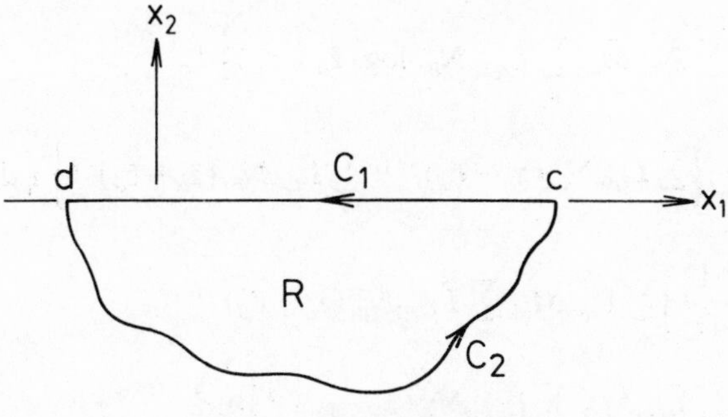

Fig. 6.4.1

where λ is defined in the previous section. Now the integral along C_1 is zero by virtue of the fact that Φ_{kj} (as defined by (6.4.1)) and ϕ_k are both zero as $x_2 = 0$. Hence (6.4.3) reduces to

$$\lambda\phi_j(\mathbf{x}_0) + F^{-1}\int_{C_2} [P_i(\mathbf{x})\Phi_{ij}(\mathbf{x}, \mathbf{x}_0) - \Gamma_{ij}(\mathbf{x}, \mathbf{x}_0)\phi_i(\mathbf{x})]\,ds(\mathbf{x}) = 0. \qquad (6.4.4)$$

If ϕ_k is specified and non-zero on C_2 then two approaches to the problem of finding ϕ_k in R are possible. Firstly the integral equation (6.4.3) may be used directly in which case it simplifies to

$$\lambda\phi_j(\mathbf{x}_0) - F^{-1}\int_{C_1} \Gamma_{ij}(\mathbf{x}, \mathbf{x}_0)\phi_i(\mathbf{x})\,ds(\mathbf{x})$$

$$+ F^{-1}\int_{C_2} [P_i(\mathbf{x})\Phi_{ij}(\mathbf{x}, \mathbf{x}_0) - \Gamma_{ij}(\mathbf{x}, \mathbf{x}_0)\phi_i(\mathbf{x})]\,ds(\mathbf{x}) = 0. \qquad (6.4.5)$$

Alternatively the method of superposition may be used. Write

$$\phi_k = \phi_k^{(1)} + \phi_k^{(2)}, \qquad (6.4.6)$$

where $\phi_k^{(1)}$ and $\phi_k^{(2)}$ are both solutions to (1.1.1). On C_1 $\phi_k^{(1)} = 0$ while $\phi_k^{(2)}$ is put equal to the specified values of ϕ_k. The solution $\phi_k^{(2)}$ is then determined from Section 3.2 while $\phi_k^{(1)}$ is determined from (6.4.4) where the boundary conditions on C_2 are suitably altered to allow for the effect of $\phi_k^{(2)}$.

Case II $P_i = 0$ on C_1. The d_{rj} are defined by (6.3.7) while the corresponding ϕ_k and Γ_{ij} are (from Section 5.4)

$$\Phi_{kj} = \frac{1}{4\pi}\left\{\sum_\alpha A_{k\alpha}N_{\alpha r}\log(z_\alpha - c_\alpha) + \sum_\alpha \bar{A}_{k\alpha}\bar{N}_{\alpha r}\log(\bar{z}_\alpha - \bar{c}_\alpha)\right\}d_{rj}$$

$$-\frac{1}{4\pi}\left\{\sum_\alpha A_{k\alpha}M_{\alpha q}\sum_\beta \bar{L}_{q2\beta}\bar{N}_{\beta r}\log(z_\alpha - \bar{c}_\beta)\right.$$

$$\left. +\sum_\alpha \bar{A}_{k\alpha}\bar{M}_{\alpha q}\sum_\beta L_{q2\beta}N_{\beta r}\log(\bar{z}_\alpha - c_\beta)\right\}d_{rj}, \qquad (6.4.7)$$

$$\Gamma_{ij} = \frac{1}{4\pi}\left\{\sum_\alpha L_{ip\alpha}N_{\alpha r}(z_\alpha - c_\alpha)^{-1} + \sum_\alpha \bar{L}_{ip\alpha}\bar{N}_{\alpha r}(\bar{z}_\alpha - \bar{c}_\alpha)^{-1}\right\}n_p d_{rj}$$

$$-\frac{1}{4\pi}\left\{\sum_\alpha L_{ip\alpha}M_{\alpha k}\sum_\beta \bar{L}_{k2\beta}\bar{N}_{\beta r}(z_\alpha - \bar{c}_\beta)^{-1}\right.$$

$$\left. +\sum_\alpha \bar{L}_{ip\alpha}\bar{M}_{\alpha k}\sum_\beta L_{k2\beta}N_{\beta r}(\bar{z}_\alpha - c_\beta)^{-1}\right\}n_p d_{rj}. \qquad (6.4.8)$$

The boundary integral equation for this problem is given by (6.4.3) where, in this case, the Φ_{kj} and Γ_{ij} are given by (6.4.7) and (6.4.8). Now $P_i = 0$ and $\Gamma_{ij} = 0$ on $x_2 = 0$ and hence the boundary integral equation reduces to (6.4.4).

If P_i is specified and non-zero on C_2 then the two possible approaches outlined for the corresponding case for ϕ_k are again applicable. If (6.4.3) is used directly with Φ_{kj} and Γ_{ij} determined by (6.4.7) and (6.4.8) then (6.4.3) reduces to

$$\lambda\phi_j(\mathbf{x}_0) + F^{-1}\int_{C_1} P_i(\mathbf{x})\Phi_{ij}(\mathbf{x}, \mathbf{x}_0)\,ds(\mathbf{x})$$

$$+ F^{-1}\int_{C_2} [P_i(\mathbf{x})\Phi_{ij}(\mathbf{x}, \mathbf{x}_0) - \Gamma_{ij}(\mathbf{x}, \mathbf{x}_0)\phi_i(\mathbf{x})]\,ds(\mathbf{x}) = 0. \qquad (6.4.9)$$

Use of the superposition procedure leads to a solution ϕ_k with corresponding P_i in the form

$$\phi_k = \phi_k^{(1)} + \phi_k^{(2)}, \qquad (6.4.10)$$

$$P_i = P_i^{(1)} + P_i^{(2)}, \qquad (6.4.11)$$

where $\phi_k^{(N)}$ (with corresponding $P_i^{(N)}$) for $N = 1, 2$ are both solutions to (1.1.1). On C_1 $P_i^{(1)} = 0$ while $P_i^{(2)}$ is put equal to the specified values of P_i. The solution $\phi_k^{(2)}$ is determined from Section 3.3 while $\phi_k^{(1)}$ is found from (6.4.4) where the boundary conditions on C_2 are suitably altered to allow for the effect of $\phi_k^{(2)}$.

6.5 A particular class of mixed boundary-value problems

A solution to (1.1.1) is required for a region R with boundary $C = C_1 + C_2$ (Fig. 6.4.1). On C_1 the following boundary conditions hold.

Problem I

$$\phi_i(x_1, 0) = 0 \quad \text{for} \quad x_1 < a \quad \text{and} \quad x_1 > b, \qquad (6.5.1)$$

$$P_i(x_1, 0) = 0 \quad \text{for} \quad a < x_1 < b. \qquad (6.5.2)$$

Problem II

$$P_i(x_1, 0) = 0 \quad \text{for} \quad x_1 < a \quad \text{and} \quad x_1 > b, \qquad (6.5.3)$$

$$\phi_i(x_1, 0) = 0 \quad \text{for} \quad a < x_1 < b. \qquad (6.5.4)$$

For Problem I let the d_{rj} be defined by (6.3.7). The corresponding Φ_{kj} and Γ_{ij} obtained, in this case, from Section 5.5 are

$$\Phi_{kj} = \frac{1}{4\pi} \left\{ \sum_\alpha A_{k\alpha} N_{\alpha r} \log (z_\alpha - c_\alpha) + \sum_\alpha \bar{A}_{k\alpha} \bar{N}_{\alpha r} \log (\bar{z}_\alpha - \bar{c}_\alpha) \right\} d_{rj}$$

$$- \frac{1}{4\pi} \left\{ \sum A_{k\alpha} N_{\alpha q} \sum_\beta \bar{A}_{q\beta} \bar{N}_{\beta r} \log (z_\alpha - \bar{c}_\beta) \right.$$

$$+ \sum_\alpha \bar{A}_{k\alpha} \bar{N}_{\alpha q} \sum_\beta A_{q\beta} N_{\beta r} \log (\bar{z}_\alpha - c_\beta) \bigg\} d_{rj}$$

$$+ \sum_\alpha A_{k\alpha} N_{\alpha r} \Theta_{rj}(z_\alpha) - \sum_\alpha \bar{A}_{k\alpha} \bar{N}_{\alpha r} \Theta_{rj}(\bar{z}_\alpha), \tag{6.5.5}$$

$$\Gamma_{ij} = \frac{1}{4\pi} \left\{ \sum_\alpha L_{ip\alpha} N_{\alpha r} (z_\alpha - c_\alpha)^{-1} + \sum_\alpha \bar{L}_{ip\alpha} \bar{N}_{\alpha r} (\bar{z}_\alpha - \bar{c}_\alpha)^{-1} \right\} n_p d_{rj}$$

$$- \frac{1}{4\pi} \left\{ \sum_\alpha L_{ip\alpha} N_{\alpha k} \sum_\beta \bar{L}_{k2\beta} \bar{N}_{\beta r} (z_\alpha - \bar{c}_\beta)^{-1} \right.$$

$$+ \sum_\alpha \bar{L}_{ip\alpha} \bar{N}_{\alpha k} \sum_\beta L_{k2\beta} N_{\beta r} (\bar{z}_\alpha - c_\beta)^{-1} \bigg\} n_p d_{rj}$$

$$+ \left\{ \sum_\alpha L_{ip\alpha} N_{\alpha k} \Theta'_{kr}(z_\alpha) - \sum_\alpha \bar{L}_{ip\alpha} \bar{N}_{\alpha k} \Theta'_{kr}(\bar{z}_\alpha) \right\} n_p d_{rj}, \tag{6.5.6}$$

where

$$\Theta'_{kj}(z) = \sum_\gamma T_{k\gamma} X_\gamma(z) \Psi_{\gamma j}(z). \tag{6.5.7}$$

In (6.5.7)

$$\Psi_{\gamma j}(z) = \frac{R_{\gamma i}}{(1 - e^{-2\pi i m})} \sum_\alpha \left\{ \frac{\mathcal{W}_{i\alpha j}}{z - c_\alpha} [Y_\gamma(z) - Y_\gamma(c_\alpha)] \right.$$

$$+ \frac{\bar{\mathcal{W}}_{i\alpha j}}{z - \bar{c}_\alpha} [Y_\gamma(z) - Y_\gamma(\bar{c}_\alpha)] \bigg\}, \tag{6.5.8}$$

and

$$X_\gamma(z) = (z - b)^{m-1} (z - a)^{-m}, \tag{6.5.9}$$

where

$$\mathcal{W}_{i\alpha j} = \frac{1}{4\pi} \{ L_{i2\alpha} N_{\alpha r} - \bar{C}_{iq} L_{q2\alpha} N_{\alpha r} \} d_{rj} \tag{6.5.10}$$

and

$$Y_\gamma(z) = X_\gamma^{-1}(z)[1 - X_\gamma(z)\{(m-1)b + ma + z\}].\qquad(6.5.11)$$

Also, $R_{\gamma i}$ and m are defined in Section 3.4 while C_{iq} is defined by (1.2.17).

The required boundary integral equation for this problem is, from (6.2.1),

$$\lambda\phi_j(\mathbf{x}_0) + F^{-1}\int_{C=C_1+C_2} [P_i(\mathbf{x})\Phi_{ij}(\mathbf{x}, \mathbf{x}_0) - \Gamma_{ij}(\mathbf{x}, \mathbf{x}_0)\phi_i(\mathbf{x})]\,ds(\mathbf{x}) = 0$$

$$(6.5.12)$$

where λ is defined in Section 6.3. The integral along C_1 is zero by virtue of the fact that, on $x_2 = 0$, Φ_{ij} (as defined by (6.5.5)) and ϕ_i are zero for $x_1 < a$ and $x_1 > b$ while Γ_{ij} and P_i are zero for $a < x_1 < b$. Hence (6.5.12) reduces to

$$\lambda\phi_j(\mathbf{x}_0) + F^{-1}\int_{C_2} [P_i(\mathbf{x})\Phi_{ij}(\mathbf{x}, \mathbf{x}_0) - \Gamma_{ij}(\mathbf{x}, \mathbf{x}_0)\phi_i(\mathbf{x})]\,ds(\mathbf{x}) = 0.\qquad(6.5.13)$$

If ϕ_i is specified and non-zero on $x_2 = 0$ for $a < x_1 < b$ and condition (6.5.2) still holds then two approaches are possible. Firstly, equation (6.5.12) may be used directly in which case it reduces to

$$\lambda\phi_j(\mathbf{x}_0) - F^{-1}\int_{C_3} \Gamma_{ij}(\mathbf{x}, \mathbf{x}_0)\phi_i(\mathbf{x})\,ds(\mathbf{x})$$

$$+ F^{-1}\int_{C_2} [P_i(\mathbf{x})\Phi_{ij}(\mathbf{x}, \mathbf{x}_0) - \Gamma_{ij}(\mathbf{x}, \mathbf{x}_0)\phi_i(\mathbf{x})]\,ds(\mathbf{x}) = 0.\qquad(6.5.14)$$

where C_3 consists of that part of the x_1-axis lying between $x_1 = a$ and $x_1 = b$. Alternatively, the method of superposition may be employed. Let

$$\phi_k = \phi_k^{(1)} + \phi_k^{(2)},\qquad(6.5.15)$$

$$P_i = P_i^{(1)} + P_i^{(2)},\qquad(6.5.16)$$

where $\phi_k^{(1)}$ and $\phi_k^{(2)}$ are both solutions to (1.1.1). On C_1 $\phi_k^{(1)} = 0$ for $a < x_1 < b$ while $P_i = 0$ for $x_1 < a$ and $x_1 > b$ while $\phi_k^{(2)}$ is put equal to the specified values of ϕ_k and $P_i^{(2)} = 0$ for $x_1 < a$ and $x_1 > b$. The solution $\phi_k^{(2)}$ is then determined from Section 3.5 while $\phi_k^{(1)}$ is determined from (6.4.4) where the boundary conditions on C_2 are suitably altered to allow for the effect of $\phi_k^{(2)}$.

For Problem II the d_{rj} are again defined by (6.3.7) while the corresponding Φ_{kj} and Γ_{ij} obtained from Section 5.5 are

$$\Phi_{kj} = \frac{1}{4\pi} \left\{ \sum_{\alpha} A_{k\alpha} N_{\alpha r} \log (z_\alpha - c_\alpha) + \sum_{\alpha} \bar{A}_{k\alpha} \bar{N}_{\alpha r} \log (\bar{z}_\alpha - \bar{c}_\alpha) \right\} d_{rj}$$

$$- \frac{1}{4\pi} \left\{ \sum_{\alpha} A_{k\alpha} M_{\alpha q} \sum_{\beta} \bar{L}_{q2\beta} \bar{N}_{\beta r} \log (z_\alpha - \bar{c}_\beta) \right.$$

$$\left. + \sum_{\alpha} \bar{A}_{k\alpha} \bar{M}_{\alpha q} \sum_{\beta} L_{q2\beta} N_{\beta r} \log (\bar{z}_\alpha - c_\beta) \right\} d_{rj}$$

$$+ \left\{ \sum_{\alpha} A_{k\alpha} M_{\alpha r} \Theta_{rj}(z_\alpha) - \sum_{\alpha} \bar{A}_{k\alpha} \bar{M}_{\alpha r} \Theta_{rj}(\bar{z}_\alpha) \right\} d_{rj} \qquad (6.5.17)$$

$$\Gamma_{ij} = \frac{1}{4\pi} \left\{ \sum_{\alpha} L_{ip\alpha} N_{\alpha r} (z_\alpha - c_\alpha)^{-1} + \sum_{\alpha} \bar{L}_{ip\alpha} \bar{N}_{\alpha r} (z_\alpha - c_\alpha)^{-1} \right\} n_p d_{rj}$$

$$- \frac{1}{4\pi} \left\{ \sum_{\alpha} L_{ip\alpha} M_{\alpha k} \sum_{\beta} \bar{L}_{k2\beta} \bar{N}_{\beta r} (z_\alpha - \bar{c}_\beta)^{-1} \right.$$

$$\left. + \sum_{\alpha} \bar{L}_{ip\alpha} \bar{M}_{\alpha k} \sum_{\beta} L_{k2\beta} N_{\beta r} (\bar{z}_\alpha - c_\beta)^{-1} \right\} n_p d_{rj}$$

$$+ \left\{ \sum_{\alpha} L_{ip\alpha} M_{\alpha k} \Theta'_{kr}(z_\alpha) - \sum_{\alpha} \bar{L}_{ip\alpha} \bar{M}_{\alpha k} \Theta'_{kr}(\bar{z}_\alpha) \right\} n_p d_{rj}, \qquad (6.5.18)$$

where

$$\Theta'_{kj}(z) = \sum_{\gamma} T_{k\gamma} X_\gamma(z) \Psi_{\gamma j}(z), \qquad (6.5.19)$$

with

$$\Psi_{\gamma j}(z) = \frac{R_{\gamma i}}{(1 - e^{-2\pi i m})} \sum_{\alpha} \left\{ \frac{\mathcal{V}_{i\alpha j}}{z - c_\alpha} [Y(z) - Y(c_\alpha)] \right.$$

$$\left. + \frac{\mathcal{V}_{i\alpha j}}{z - \bar{c}_\alpha} [Y(z) - Y(\bar{c}_\alpha)] \right\}. \qquad (6.5.20)$$

In (6.5.20) $X_\gamma(z)$ and $Y(z)$ are defined by (6.5.9) and (6.5.11), respectively, while

$$\mathcal{V}_{i\alpha j} = \frac{1}{4\pi} \{ A_{i\alpha} N_{\alpha r} - \bar{B}_{iq} L_{q2\alpha} N_{\alpha r} \} d_{rj}. \qquad (6.5.21)$$

The boundary integral equation for this problem is just (6.5.12). In this case the integral along C_1 is zero by virtue of the fact that, on $x_2 = 0$, Γ_{ij}

(as defined by (6.5.18)) and P_i are zero for $x_1 < a$ and $x_1 > b$ while Φ_{ij} and ϕ_i are zero for $a < x_1 < b$. Hence the boundary integral equation reduces to (6.5.13).

If P_i is specified and non-zero on $x_2 = 0$ for $a < x_1 < b$ and condition (6.5.4) still holds then the same two approaches as were outlined for Problem I are again possible for Problem II. If equation (6.5.12) is used directly then it reduces to

$$\lambda \phi_j(\mathbf{x}_0) + F^{-1} \int_{C_3} P_i(\mathbf{x}) \Phi_{ij}(\mathbf{x}, \mathbf{x}_0) \, ds(\mathbf{x})$$

$$+ F^{-1} \int_{C_2} [P_i(\mathbf{x}) \Phi_{ij}(\mathbf{x}, \mathbf{x}_0) - \Gamma_{ij}(\mathbf{x}, \mathbf{x}_0) \phi_i(\mathbf{x})] \, ds(\mathbf{x}) = 0. \qquad (6.5.22)$$

Alternatively, the method of superposition may be used in a similar way to that outlined for Problem I.

6.6   The boundary integral equation for a region with a cut

A boundary integral equation is required for the solution of (1.1.1) for a region R with boundary $C = C_1 + C_2$ (Fig. 6.6.1). The section C_1 of the boundary consists of the cut along the x_1-axis from $x_1 = -1$ to $x_1 = 1$. On C_1 $P_i = 0$ while on C_2 either the dependent variables ϕ_k are specified or the P_i are specified.

To obtain the required boundary integral equation let the d_{rj} be defined by (6.3.7). The corresponding Φ_{kj} and Γ_{ij} obtained, in this case, from

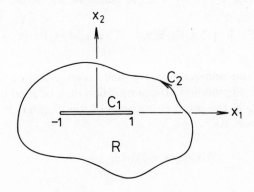

Fig. 6.6.1

Section 5.6 are

$$
\Phi_{kj} = \frac{1}{4\pi} \left\{ \sum_\alpha A_{k\alpha} N_{\alpha r} \log (z_\alpha - c_\alpha) + \sum_\alpha \bar{A}_{k\alpha} \bar{N}_{\alpha r} \log (\bar{z}_\alpha - \bar{c}_\alpha) \right\} d_{rj}
$$

$$
+ \frac{1}{8\pi} \left\{ A_{k\alpha} M_{\alpha p} \left[\sum_\beta L_{p2\beta} N_{\beta r} \Omega(z_\alpha, c_\beta) + \sum_\beta \bar{L}_{p2\beta} \bar{N}_{\beta r} \Omega(z_\alpha, \bar{c}_\beta) \right] \right.
$$

$$
\left. + \bar{A}_{k\alpha} \bar{M}_{\alpha p} \left[\sum_\beta \bar{L}_{p2\beta} \bar{N}_{\beta r} \bar{\Omega}(\bar{z}_\alpha, \bar{c}_\beta) + \sum_\beta L_{p2\beta} N_{\beta r} \bar{\Omega}(\bar{z}_\alpha, c_\beta) \right] \right\} d_{rj}.
$$

$$(6.6.1)$$

$$
\Gamma_{ij} = \frac{1}{4\pi} \left\{ \sum_\alpha L_{ip\alpha} N_{\alpha r} (z_\alpha - c_\alpha)^{-1} + \sum_\alpha \bar{L}_{ip\alpha} \bar{N}_{\alpha r} (\bar{z}_\alpha - \bar{c}_\alpha)^{-1} \right\} n_p d_{rj}
$$

$$
- \frac{1}{8\pi} \left\{ \sum_\alpha L_{ip\alpha} M_{\alpha k} (z_\alpha^2 - 1)^{-1/2} \sum_\beta [L_{k2\beta} N_{\beta r} \chi(z_\alpha, c_\beta) \right.
$$

$$
+ \bar{L}_{k2\beta} \bar{N}_{\beta r} \chi(z_\alpha, \bar{c}_\beta)]
$$

$$
+ \sum_\alpha \bar{L}_{ip\alpha} \bar{M}_{\alpha k} (\bar{z}_\alpha^2 - 1)^{-\frac{1}{2}} \sum_\beta [\bar{L}_{k2\beta} \bar{N}_{\beta r} \chi(\bar{z}_\alpha, \bar{c}_\beta)
$$

$$
\left. + L_{k2\beta} N_{\beta r} \bar{\chi}(\bar{z}_\alpha, c_\beta)] \right\} n_p d_{rj}.
$$

$$(6.6.2)$$

where $\Omega(z, c)$ and $\chi(z, c)$ are defined, respectively, by (5.6.13) and (5.6.16). From (6.2.1) the boundary integral equation is

$$
\lambda \phi_j(\mathbf{x}_0) + F^{-1} \int_{C = C_1 + C_2} [P_i(\mathbf{x}) \Phi_{ij}(\mathbf{x}, \mathbf{x}_0) - \Gamma_{ij}(\mathbf{x}, \mathbf{x}_0) \phi_i(\mathbf{x})] \, ds(\mathbf{x}) = 0,
$$

$$(6.6.3)$$

where λ is defined in Section 6.3.

Now $P_i = 0$ and $\Gamma_{ij} = 0$ on C_1 and hence (6.6.3) reduces to

$$
\lambda \phi_j(\mathbf{x}_0) + F^{-1} \int_{C_2} [P_i(\mathbf{x}) \Phi_{ij}(\mathbf{x}, \mathbf{x}_0) - \Gamma_{ij}(\mathbf{x}, \mathbf{x}_0) \phi_i(\mathbf{x})] \, ds(\mathbf{x}) = 0. \qquad (6.6.4)
$$

If P_i is given and non-zero on C_1 then either equation (6.6.3) may be used directly or the method of superposition may be used in a similar way to that outlined in the previous sections of this chapter. If (6.6.3) is used directly it reduces to

$$
\lambda \phi_j(\mathbf{x}_0) + F^{-1} \int_{C_1} P_i(\mathbf{x}) \Phi_{ij}(\mathbf{x}, \mathbf{x}_0) \, ds(\mathbf{x})
$$

$$
+ F^{-1} \int_{C_2} [P_i(\mathbf{x}) \Phi_{ij}(\mathbf{x}, \mathbf{x}_0) - \Gamma_{ij}(\mathbf{x}, \mathbf{x}_0) \phi_i(\mathbf{x})] \, ds(\mathbf{x}) = 0. \qquad (6.6.5)
$$

6.7 The boundary integral equation for a strip

A boundary integral equation is required for the solution of (1.1.1) for a region R with boundary $C = \sum_{i=1}^{4} C_i$ (Fig. 6.7.1). The sections C_1 and C_3 consist of the lines $x_2 = h$ and $x_2 = -h$ respectively while C_2 and C_4 consist of the lines joining C_1 to C_3. On C_1 and C_3 $P_i = 0$ while on C_2 and C_4 either ϕ_i or P_i are specified.

To obtain the boundary integral equation for this problem let the d_{rj} be defined by (6.3.7). The corresponding Φ_{kj} and Γ_{ij} obtained from Section 5.7 are

$$\Phi_{kj} = \frac{1}{4\pi} \left\{ \sum_\alpha A_{k\alpha} N_{\alpha r} \log (z_\alpha - c_\alpha) + \sum_\alpha \bar{A}_{k\alpha} \bar{N}_{\alpha j} \log (\bar{z}_\alpha - \bar{c}_\alpha) \right\} d_{rj}$$

$$+ \sum_\alpha A_{k\alpha} \Theta_{\alpha j}(z_\alpha) + \sum_\alpha \bar{A}_{k\alpha} \bar{\Theta}_{\alpha j}(\bar{z}_\alpha), \qquad (6.7.1)$$

$$\Gamma_{ij} = \frac{1}{4\pi} \left\{ \sum_\alpha L_{ip\alpha} N_{\alpha k}(z_\alpha - c_\alpha)^{-1} + \sum_\alpha \bar{L}_{ip\alpha} \bar{N}_{\alpha k}(\bar{z}_\alpha - \bar{c}_\alpha)^{-1} \right\} n_p d_{kj}$$

$$+ \left\{ \sum_\alpha L_{ip\alpha} \Theta'_{\alpha j}(z) + \sum_\alpha \bar{L}_{ip\alpha} \bar{\Theta}'_{\alpha j}(\bar{z}_\alpha) \right\} n_p, \qquad (6.7.2)$$

where

$$\Theta_{\alpha j}(z) = \frac{1}{2\pi} \int_0^\infty \{ E_{\alpha j}(p) \exp (ipz_\alpha) + F_{\alpha j}(p) \exp (-ipz_\alpha) \} \, dp. \qquad (6.7.3)$$

In (6.7.3) the $E_{\alpha j}$ and $F_{\alpha j}$ are obtained from the equations

$$\sum_\alpha [S_{i\alpha} E_{\alpha j} + \bar{R}_{i\alpha} \bar{F}_{\alpha j}] = (2p)^{-1} d_{kj} \sum_\alpha L_{i2\alpha} N_{\alpha k} \exp [ip(\tau_\alpha h - c_\mathbf{a})], \qquad (6.7.4)$$

$$\sum_\alpha [R_{i\alpha} E_{\alpha j} + \bar{S}_{i\alpha} \bar{F}_{\alpha j}] = (2p)^{-1} d_{kj} \sum_\alpha \bar{L}_{i2\alpha} \bar{N}_{\alpha k} \exp [-ip(\bar{\tau}_{\alpha h} + c_\alpha)]. \qquad (6.7.5)$$

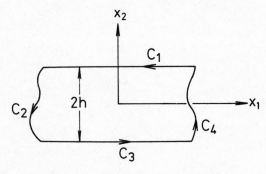

Fig. 6.7.1

From (6.2.1) the boundary integral equation for this problem is

$$\lambda \phi_j(\mathbf{x}_0) + F^{-1} \int_{C = \sum_1^4 C_i} [P_i(\mathbf{x}) \Phi_{ij}(\mathbf{x}, \mathbf{x}_0) - \Gamma_{ij}(\mathbf{x}, \mathbf{x}_0) \phi_i(\mathbf{x})] \, ds(\mathbf{x}) = 0 \quad (6.7.6)$$

where λ is defined in Section 6.3. Since P_i and Γ_{ij} are both zero on C_1 and C_3 this equation reduces to

$$\lambda \phi_j(\mathbf{x}_0) + F^{-1} \int_{C_2 + C_4} [P_i(\mathbf{x}) \Phi_{ij}(\mathbf{x}, \mathbf{x}_0) - \Gamma_{ij}(\mathbf{x}, \mathbf{x}_0) \phi_i(\mathbf{x})] \, ds(\mathbf{x}) = 0. \quad (6.7.7)$$

If P_i is given and non-zero on C_1 and C_3 then the method of superposition may be used in a similar manner to that described for some of the previous problems considered in the chapter. Alternatively, (6.7.6) may be used directly in which case it reduces to

$$\lambda \phi_j(\mathbf{x}_0) + F^{-1} \int_{C_1 + C_3} P_i(\mathbf{x}) \Phi_{ij}(\mathbf{x}, \mathbf{x}_0) \, ds(\mathbf{x})$$

$$+ F^{-1} \int_{C_2 + C_4} [P_i(\mathbf{x}) \Phi_{ij}(\mathbf{x}, \mathbf{x}_0) - \Gamma_{ij}(\mathbf{x}, \mathbf{x}_0) \phi_i(\mathbf{x})] \, ds(\mathbf{x}) = 0. \quad (6.7.8)$$

7

Applications of the boundary integral equations

7.1 Introduction

The boundary integral equations derived previously are used in this chapter to obtain the numerical solution of some particular boundary-value problems. The numerical procedure is described in some detail and, whenever possible, the numerical solution is compared with the analytical solution in order to obtain some indication of the accuracy of the numerical method. Many of the problems considered only involve a single second order elliptic partial differential equation. However, the numerical procedures outlined for the single equation are readily extended to systems of two or more equations and some particular problems for a system of two equations are discussed in Section 7.6.

7.2 Some boundary-value problems for Laplace's equation

The boundary-value problem to be considered is to find a solution to

$$\frac{\partial^2 \phi}{\partial x_1^2} + \frac{\partial^2 \phi}{\partial x_2^2} = 0 \tag{7.2.1}$$

which is valid in a region R in E^2 with boundary C.

On C either ϕ is specified or the normal derivative

$$\frac{\partial \phi}{\partial n} = \frac{\partial \phi}{\partial x_1} n_1 + \frac{\partial \phi}{\partial x_2} n_2 \tag{7.2.2}$$

is given with n_i the unit outward normal to R. If the normal derivative is given at all points on C then it cannot be specified completely arbitrarily but is constrained by the condition (1.1.6). In this case (1.1.6) yields

$$\int_C \frac{\partial \phi}{\partial n} \, ds = 0. \tag{7.2.3}$$

The boundary integral equation for the solution of this problem is (6.3.4) with Φ_{ij} and Γ_{ij} given by (6.3.5)–(6.3.7) where the constants occurring in these equations are given by (2.2.3). Using (2.2.3) with $K = 1$ in (6.3.4)–(6.3.7) and dropping the subscripts which are unnecessary in this case it follows that (6.3.4) may be written (with the arbitrary constant $F = 1$)

$$\lambda\phi(\mathbf{x}_0) + \int_C [P(\mathbf{x})\Phi(\mathbf{x}, \mathbf{x}_0) - \Gamma(\mathbf{x}, \mathbf{x}_0)\phi(\mathbf{x})]\, ds(\mathbf{x}) = 0, \tag{7.2.4}$$

where

$$\Phi = \frac{1}{4\pi}\{\log(z - c) + \log(\bar{z} - \bar{c})\}$$

$$= \frac{1}{2\pi}\log|z - c|$$

$$= \frac{1}{4\pi}\log[(x_1 - a)^2 + (x_2 - b)^2], \tag{7.2.5}$$

$$\Gamma = \frac{1}{4\pi}\left\{\frac{1}{z - c} + \frac{1}{\bar{z} - \bar{c}}\right\}n_1 + \frac{i}{4\pi}\left\{\frac{1}{z - c} - \frac{1}{\bar{z} - \bar{c}}\right\}n_2$$

$$= \frac{1}{2\pi}\left[\frac{(x_1 - a)n_1 + (x_2 - b)n_2}{(x_1 - a)^2 + (x_2 - b)^2}\right] \tag{7.2.6}$$

and

$$P = \frac{\partial\phi}{\partial x_1}n_1 + \frac{\partial\phi}{\partial x_2}n_2.$$

As a test problem consider the case when the domain R consists of the square $ABCD$ (Fig. 7.2.1). The boundary conditions are

$$\begin{aligned}
\phi(0.6, x_2) &= 0.6 + x_2 &&\text{on}&& AB, \\
\phi(x_1, 0.6) &= x_1 + 0.6 &&\text{on}&& BC, \\
\phi(-0.6, x_2) &= -0.6 + x_2 &&\text{on}&& CD, \\
\phi(x_1, -0.6) &= x_1 - 0.6 &&\text{on}&& DA.
\end{aligned} \tag{7.2.7}$$

This problem admits the simple analytical solution $\phi = x_1 + x_2$.

Equation (7.2.4) may be used to determine numerical values of P on the boundary C as follows. In (7.2.4) let $\phi = 1$ so that $P = 0$ and hence

$$\lambda = \int_C \Gamma(\mathbf{x}, \mathbf{x}_0)\, ds(\mathbf{x}). \tag{7.2.8}$$

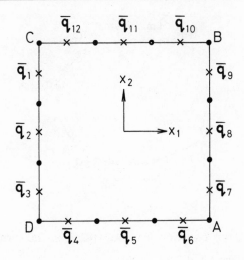

Fig. 7.2.1

Substitution of (7.2.8) into (7.2.4) yields

$$\int_C P(\mathbf{x})\Phi(\mathbf{x}, \mathbf{x}_0)\,ds(\mathbf{x}) = \int_C \Gamma(\mathbf{x}, \mathbf{x}_0)[\phi(\mathbf{x}) - \phi(\mathbf{x}_0)]\,ds(\mathbf{x}). \qquad (7.2.9)$$

The numerical technique used to solve equation (7.2.9) consists of replacing integration by summation so that a linear system of algebraic equations is obtained. This system may be solved by employing standard techniques.

Following Jaswon and Ponter [24] and Symm [39] the boundary C is divided into N segments between the boundary points with position vectors $\mathbf{q}_{m-1}$ and $\mathbf{q}_m$ for $m = 1, 2, \ldots, N$ with $\mathbf{q}_0 = \mathbf{q}_N$. P is assumed to be constant on each of these segments and to take the value P_m on the segment of the boundary lying between $\mathbf{q}_{m-1}$ and $\mathbf{q}_m$. Thus (7.2.9) may be approximated by

$$\sum_{m=1}^{N} P_m \int_{\mathbf{q}_{m-1}}^{\mathbf{q}_m} \Phi(\mathbf{x}, \mathbf{x}_0)\,ds(\mathbf{x}) = \sum_{m=1}^{N} \int_{\mathbf{q}_{m-1}}^{\mathbf{q}_m} \Gamma(\mathbf{x}, \mathbf{x}_0)[\phi(\mathbf{x}) - \phi(\mathbf{x}_0)]\,ds(\mathbf{x}).$$

$$(7.2.10)$$

It is convenient to consider a mid-point $\bar{\mathbf{q}}_m$ in each segment of the boundary $\mathbf{q}_{m-1}$ to $\mathbf{q}_m$ as shown in Fig. 7.2.2. If $\mathbf{x}_0$ is, in turn, taken to coincide with the mid-points $\bar{q}_j$ for $j = 1, 2, \ldots, N$ then equation (7.2.10) yields the N equations

$$\sum_{m=1}^{N} P_m \int_{\mathbf{q}_{m-1}}^{\mathbf{q}_m} \Phi(\mathbf{x}, \bar{\mathbf{q}}_j)\,ds(\mathbf{x}) = \sum_{m=1}^{N} \int_{\mathbf{q}_{m-1}}^{\mathbf{q}_m} \Gamma(\mathbf{x}, \bar{\mathbf{q}}_j)[\phi(\mathbf{x}) - \phi(\bar{\mathbf{q}}_j)]\,ds(\mathbf{x})$$

$$\text{for} \quad j = 1, 2, \ldots, N. \quad (7.2.11)$$

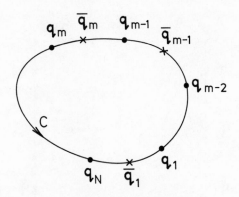

Fig. 7.2.2

More compactly, this equation may be written in the form

$$P_m a_{mj} = R_j \quad \text{for} \quad j = 1, 2, \ldots, N, \tag{7.2.12}$$

where

$$a_{mj} = \int_{\mathbf{q}_{m-1}}^{\mathbf{q}_m} \Phi(\mathbf{x}, \bar{\mathbf{q}}_j) \, ds(\mathbf{x}) \tag{7.2.13}$$

and

$$R_j = \sum_{m=1}^{N} \int_{\mathbf{q}_{m-1}}^{\mathbf{q}_m} \Gamma(\mathbf{x}, \bar{\mathbf{q}}_j)[\phi(\mathbf{x}) - \phi(\bar{\mathbf{q}}_j)] \, ds(\mathbf{x}). \tag{7.2.14}$$

Since a_{mj} and R_j may be calculated from the above integrals it follows that the only unknowns in the system of N linear equations (7.2.12) are $P_1, P_2, \ldots, P_N$. The system may thus be used to solve for the unknown P_m.

Some care is needed in evaluating the a_{mj} through (7.2.13) since $\Phi(\mathbf{x}, \mathbf{x}_0)$ has a logarithmic singularity at $\mathbf{x} = \mathbf{x}_0$. It is convenient to consider the two cases $j = m$ and $j \neq m$ separately. Firstly $j = m$ means that $\bar{\mathbf{q}}_j$ is on the path of integration in (7.2.13). By treating the integrals along the straight lines from $\mathbf{q}_{m-1}$ to $\bar{\mathbf{q}}_m$ and $\bar{\mathbf{q}}_m$ to $\mathbf{q}_m$ separately and integrating by parts it is possible to show that

$$2\pi a_{mm} = |\bar{\mathbf{q}}_m - \mathbf{q}_{m-1}| \{-1 + \log |\bar{\mathbf{q}}_m - \mathbf{q}_{m-1}|\}$$
$$+ |\mathbf{q}_m - \bar{\mathbf{q}}_m| \{-1 + \log |\mathbf{q}_m - \bar{\mathbf{q}}_m|\}. \tag{7.2.15}$$

If $j \neq m$, that is $\bar{\mathbf{q}}_j$ is not on the segment from $\mathbf{q}_{m-1}$ to $\mathbf{q}_m$, then Simpson's rule may be used to yield

$$2\pi a_{mj} = \frac{h_m}{3} \{\log |\mathbf{q}_{m-1} - \bar{\mathbf{q}}_j| + 4 \log |\mathbf{q}_m - \bar{\mathbf{q}}_j| + \log |\mathbf{q}_m - \bar{\mathbf{q}}_j|\}, \tag{7.2.16}$$

where h_m is the length of the segment from $\bar{\mathbf{q}}_m$ to $\mathbf{q}_m$.

The integrals on the right hand side of (7.2.10) have integrands which are known and finite along the paths of integration and hence may be evaluated by using a standard quadrature formula. From a practical point of view it is perhaps best to avoid the point $\mathbf{x} = \mathbf{x}_0$ in any quadrature scheme since $\Gamma(\mathbf{x}, \mathbf{x}_0)$ is singular at this point although the complete integrand is finite. Thus, it is convenient to use Simpson's $\frac{3}{8}$ rule which is a four point quadrature formula (see Abramowitz and Stegun [1]) so that the mid-point of the interval is not encountered when calculating a value for the integral.

For the purposes of the present calculations each side of the square is divided into three equal segments. In Fig. 7.2.1 the end points of each segment are indicated by dots, while the mid-points are labelled with the relevant position vector $\bar{\mathbf{q}}_j$.

Values of P_m for $m = 1, 2, \ldots, 12$ obtained from (7.2.12) are given in Table 7.2.1. These are compared with those obtained from the analytical solution. Specifically, the analytical solution is $\phi = x_1 + x_2$ so that the analytical expression for P is

$$P = n_1 + n_2. \tag{7.2.17}$$

The values of P obtained from (7.2.12) differ from those obtained from the analytical solution (7.2.17) by less than 2%. These results are of course rather more accurate than would be obtained in general since the assumption that P is constant over each segment does not introduce any error in this particular problem. Also, since the region under consideration has a boundary which consists of four straight lines there is no need, in this case, to introduce an approximation to the geometry of the boundary. This is another factor which leads to a somewhat smaller error than would generally be the case.

Table 7.2.1

Point	$P(\mathbf{x})$ analytical	$P(\mathbf{x})$ numerical	% Error
$\mathbf{q}_1$	-1	-0.987	1.3
$\mathbf{q}_2$	-1	-0.998	0.2
$\mathbf{q}_3$	-1	-1.005	0.5
$\mathbf{q}_4$	-1	-1.005	0.5
$\mathbf{q}_5$	-1	-0.998	0.2
$\mathbf{q}_6$	-1	-0.987	1.3
$\mathbf{q}_7$	1	0.987	1.3
$\mathbf{q}_8$	1	0.998	0.2
$\mathbf{q}_9$	1	1.005	0.5
$\mathbf{q}_{10}$	1	1.005	0.5
$\mathbf{q}_{11}$	1	0.998	0.2
$\mathbf{q}_{12}$	1	0.987	1.3

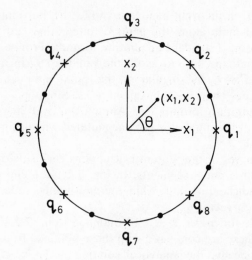

Fig. 7.2.3

As a second example consider the case when the domain R consists of the unit circle (Fig. 7.2.3). The boundary conditions are

$$T = \cos \theta + \sin \theta.$$

This problem admits the analytical solution $T = x_1 + x_2 = r(\cos \theta + \sin \theta)$, so P is again given by (7.2.17).

For the purposes of using (7.2.11) to calculate P the circumference of the circle is divided into N equal segments (the case $N = 8$ is illustrated in Fig. 7.2.3 with the dots indicating the end points of the intervals and the crosses the mid-points). With $N = 20$ equation (7.2.11) yields values of P which differ from the analytical values by a maximum of approximately 10%. For $n = 40$ the numerical and analytical value for some sample boundary points are given in Table 7.2.2. It is apparent from the table

Table 7.2.2

Boundary point $(\cos \theta, \sin \theta)$	Analytical	Numerical	% Error
$\theta = 270°$	−1.0	−0.9994	0.06
$\theta = 279°$	−0.8313	−0.8307	0.06
$\theta = 288°$	−0.6420	−0.6417	0.03
$\theta = 297°$	−0.4370	−0.4367	0.03
$\theta = 306°$	−0.2212	−0.2211	0.01
$\theta = 315°$	0	0	0
$\theta = 324°$	0.2212	0.2211	0.01
$\theta = 333°$	0.4370	0.4367	0.03
$\theta = 342°$	0.6420	0.6417	0.03
$\theta = 351°$	0.8313	0.8307	0.06
$\theta = 360°$	1.0	0.9994	0.06

that excellent results are possible provided sufficient boundary segments are taken.

7.3   A particular class of problems for Laplace's equation

In this section the boundary integral equations derived in Section 6.4 are used to consider some particular examples of a class of problems governed by Laplace's equations. The problems to be considered are such that the boundary geometry is of the general form illustrated in Fig. 6.4.1. Hence, a solution to Laplace's equation (7.2.1) is required which is valid in a region R in E^2 with boundary $C = C_1 + C_2$ where C_1 is a segment of the $0x_1$ axis. On C_1 either $\phi = 0$ or the normal derivative $P = \partial \phi / \partial n = 0$ while on C_2 either ϕ or P are specified. (More general boundary conditions on C_1 may readily be accommodated by employing the techniques outlined in Section 6.4.) If P is specified over both C_1 and C_2 then the condition (7.2.3) must be satisfied.

For illustrative purposes it will be sufficient to consider only the case when $\phi = 0$ on C_1. The boundary integral equation for the solution of this class of problems is then (6.4.4) with Φ_{ij} and Γ_{ij} given by (6.4.1), (6.4.2) and (6.3.7). Using (2.2.3) with $K = 1$ in (6.4.1), (6.4.2), (6.4.4), (6.3.7) (with $F = 1$) and dropping the subscripts it follows that

$$\lambda \phi(\mathbf{x}_0) + \int_{C_2} [P(\mathbf{x})\Phi(\mathbf{x}, \mathbf{x}_0) - \Gamma(\mathbf{x}, \mathbf{x}_0)\phi(\mathbf{x})]\, ds(\mathbf{x}) = 0, \tag{7.3.1}$$

where

$$\Phi = \frac{1}{2\pi}[\log|z - c| - \log|z - \bar{c}|]$$

$$= \frac{1}{4\pi}\{\log[(x_1 - a)^2 + (x_2 - b)^2] - \log[(x_1 - a)^2 + (x_2 + b)^2], \tag{7.3.2}$$

$$\Gamma = \frac{\partial \Phi}{\partial x_1}n_1 + \frac{\partial \Phi}{\partial x_2}n_2$$

$$= \frac{1}{2\pi}\left\{\frac{(x_1 - a)n_1 + (x_2 + b)n_2}{(x_1 - a)^2 + (x_2 - b)^2} - \frac{(x_1 - a)n_1 + (x_2 + b)n_2}{(x_1 - a)^2 + (x_2 + b)^2}\right\}. \tag{7.3.3}$$

Consider now two particular boundary-value problems for which $\phi = 0$ on C_1 and ϕ is given and non-zero on C_2. Equations (7.3.1)–(7.3.3) are used to determine P on C and, for comparison, equations (7.2.4)–(7.2.6) are also to be used for this purpose. The problems to be considered are for the square shown in Fig. 7.3.1 and the boundary data is as follows.

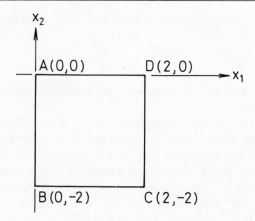

Fig. 7.3.1

Problem 1: •

$$\left.\begin{aligned}\phi &= 0 \quad && \text{on} \quad x_2 = 0\\ \phi &= -2 \quad && \text{on} \quad x_2 = -2\end{aligned}\right\} \quad \text{for} \quad 0 < x_1 < 2,$$

$$\left.\begin{aligned}\phi &= x_2 \quad && \text{on} \quad x_1 = 0\\ \phi &= x_2 \quad && \text{on} \quad x_1 = 2\end{aligned}\right\} \quad \text{for} \quad 0 < x_2 < -2.$$

(7.3.4)

Problem 2:

$$\left.\begin{aligned}\phi &= 0 \quad && \text{on} \quad x_2 = 0\\ \phi &= e^{-x_1} \quad && \text{on} \quad x_2 = -2\end{aligned}\right\} \quad \text{for} \quad 0 < x_1 < 2,$$

$$\left.\begin{aligned}\phi &= -x_2/2 \quad && \text{on} \quad x_1 = 0\\ \phi &= -x_2 e^{-2}/2 \quad && \text{on} \quad x_1 = 2\end{aligned}\right\} \quad \text{for} \quad 0 < x_2 < -2.$$

(7.3.5)

The first of these problems admits the simple analytical solution $\phi = x_2$ while the second apparently has no simple analytical solution.

The procedure for using (7.3.1)–(7.3.3) for determining P numerically on C is similar to the method outlined in the previous section. First let $\phi = 1$ in (6.4.4) to obtain

$$\lambda = \int_C \Gamma(\mathbf{x}, \mathbf{x}_0) \, ds(\mathbf{x}) \tag{7.3.6}$$

$$= \int_{C_1} \Gamma(\mathbf{x}, \mathbf{x}_0) \, ds(\mathbf{x}) + \int_{C_2} \Gamma(\mathbf{x}, \mathbf{x}_0) \, ds(\mathbf{x})$$

and using (7.3.3)

$$\lambda = \int_{C_2} \Gamma(\mathbf{x}, \mathbf{x}_0) \, ds(\mathbf{x}) - \frac{b}{\pi} \int_c^d \frac{dx_1}{(x_1 - a)^2 + b^2}$$

$$= \int_{C_2} \Gamma(\mathbf{x}, \mathbf{x}_0) \, ds(\mathbf{x}) - \frac{1}{\pi} \left\{ \tan^{-1}\left(\frac{d-a}{b}\right) - \tan^{-1}\left(\frac{c-a}{b}\right) \right\}, \tag{7.3.7}$$

where c and d are shown in Fig. 6.4.1. Substitution of this expression for λ into (7.3.1) yields

$$\int_{C_2} P(\mathbf{x})\Phi(\mathbf{x}, \mathbf{x}_0)\, ds(\mathbf{x}) = \int_{C_2} \Gamma(\mathbf{x}, \mathbf{x}_0)[\phi(\mathbf{x}) - \phi(\mathbf{x}_0)]\, ds(\mathbf{x})$$

$$+ \phi(\mathbf{x}_0)\frac{1}{\pi}\left\{\tan^{-1}\left(\frac{d-a}{b}\right) - \tan^{-1}\left(\frac{c-a}{b}\right)\right\}. \quad (7.3.8)$$

This equation may be solved numerically by using a similar procedure to the one used in the previous section for equation (7.2.9). The boundary C_2 is divided into N segments between $\mathbf{q}_{m-1}$ and $\mathbf{q}_m$ for $m = 1, 2, \ldots, N$ but in this case $\mathbf{q}_0$ and $\mathbf{q}_N$ are the position vectors of the ends c and d, respectively, of C_1.

By defining P_m and $\bar{\mathbf{q}}_j$ as before equation (7.3.8) becomes

$$\sum_{i=1}^{N} P_N \int_{\mathbf{q}_{m-1}}^{\mathbf{q}_m} \Phi(\mathbf{x}, \bar{\mathbf{q}}_j)\, ds(\mathbf{x}) = \sum_{m=1}^{N} \int_{\mathbf{q}_{m-1}}^{\mathbf{q}_m} \Gamma(\mathbf{x}, \bar{\mathbf{q}}_j)[\Phi(\mathbf{x}) - \Phi(\bar{\mathbf{q}}_j)]\, ds(\mathbf{x})$$

$$+ \Phi(\bar{\mathbf{q}}_j)\frac{1}{\pi}\left\{\tan^{-1}\left(\frac{d-a_j}{b_j}\right) - \tan^{-1}\left(\frac{c-a_j}{b_j}\right)\right\}$$

$$\text{for} \quad j = 1, 2, \ldots, N.$$

$$(7.3.9)$$

where $\bar{\mathbf{q}}_j = (a_j, b_j)$. This equation can be written in the form (7.2.12) with a_{mj} again defined by (7.2.13) but with the right hand side R_j given by

$$R_j = \sum_{m=1}^{N} \int_{\mathbf{q}_{m-1}}^{\mathbf{q}_m} \Gamma(\mathbf{x}, \bar{\mathbf{q}}_j)[\phi(\mathbf{x}) - \phi(\bar{\mathbf{q}}_j)]\, ds(\mathbf{x})$$

$$+ \phi(\bar{\mathbf{q}}_j)\frac{1}{\pi}\left\{\tan^{-1}\left(\frac{d-a_j}{b_j}\right) - \tan^{-1}\left(\frac{c-a_j}{b_j}\right)\right\}. \quad (7.3.10)$$

Although a_{mj} is given by (7.2.13) the expression for Φ in that equation is now given by (7.3.2) in place of (7.2.5). However, the first term in (7.3.2) is identical to (7.2.5) and hence the integral involving this term may be evaluated by using the procedure described in the previous section. The second term in (7.3.2) is analytic on the path C_2 of integration and hence the integral may be readily evaluated by using a simple quadrature formula such as Simpson's rule. Also, as in the previous section, the integral in (7.3.10) is everywhere finite and may be conveniently evaluated by employing Simpson's $\frac{3}{8}$ rule.

Both Problems 1 and 2 have $\phi = 0$ on $x_2 = 0$ and both the solution defined by (7.2.4)–(7.2.6) and the solution defined by (7.2.1)–(7.2.3) are applicable. As Problem 1 has the analytical solution $\phi = x_2$ it follows that

$P = \partial\phi/\partial n$ is given by

$$\left.\begin{array}{llll} P = 0 & \text{on} & x_2 = 0 \\ P = 0 & \text{on} & x_2 = -2 \end{array}\right\} \quad \text{for} \quad 0 < x_1 < 2,$$

$$\left.\begin{array}{llll} P = -1 & \text{on} & x_1 = 0 \\ P = 1 & \text{on} & x_1 = 2 \end{array}\right\} \quad \text{for} \quad 0 < x_2 < -2.$$

(7.3.11)

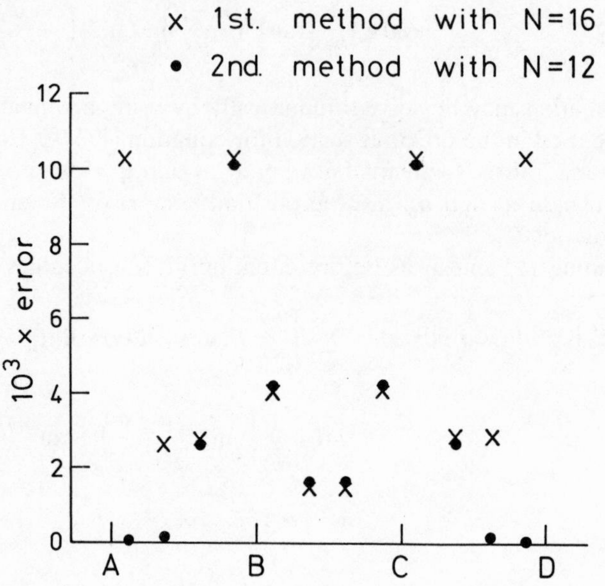

x   1st. method with N = 16

•   2nd. method with N = 12

Fig. 7.3.2

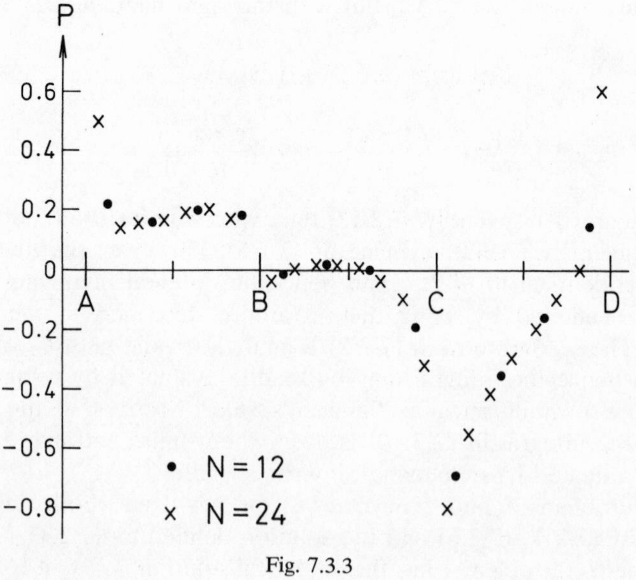

Fig. 7.3.3

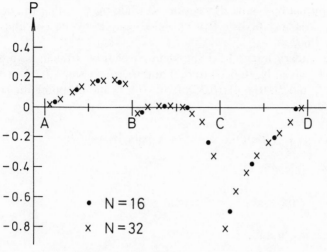

Fig. 7.3.4

Both the method of this section and the method of the previous section give numerical values of P which agree very closely with those generated by (7.3.11). Figure 7.3.2 shows the absolute value of the difference between the analytical and numerical results. The points plotted show that when the same number of intervals (four) were used on each side of the boundary the maximum error was about the same for both methods. However, the first method involved inverting a 16×16 matrix while the second method involved a 12×12 matrix since the integral was not taken round the whole boundary.

Problem 2 does not admit a simple analytical solution so it is not possible to determine the precise error involved in using the numerical procedures. However, as the step size decreases both methods converge rapidly and give good results for as few as four segments on each side. (see Figs 7.3.3 and 7.3.4). The only additional information obtained by taking more segments is that the jumps in P can be drawn more accurately. It is apparent from Figs 7.3.3 and 7.3.4 that the agreement between the results for the two methods is good except at the end points A and D.

7.4 Problems for a region with a cut

A solution to Laplace's equation (7.2.1) is required which is valid in a region R with boundary $C = C_1 + C_2$ as shown in Fig. 6.6.1. The section C_1 of the boundary consists of the cut taken along the x_1-axis from $x_1 = -1$ to $x_1 = 1$ while C_2 is a general contour surrounding the cut. On

C_1 the normal derivative $P = \partial\phi/\partial n = 0$ while on C_2 either the dependent variable ϕ is specified or P is specified subject to the condition (7.2.3) being satisfied.

The boundary integral for the solution of this problem is (6.6.4) with Φ_{ij} and Γ_{ij} given by (6.6.1), (6.6.2) and (6.3.7). Using (2.2.3) with $K = 1$ in (6.6.1), (6.6.2), (6.6.4), (6.3.7) (with $F = 1$) and dropping the subscripts it follows that

$$\lambda\phi(\mathbf{x}_0) + \int_{C_2} [P(\mathbf{x})\Phi(\mathbf{x}, \mathbf{x}_0) - \Gamma(\mathbf{x}, \mathbf{x}_0)\phi(\mathbf{x})] \, ds(\mathbf{x}) = 0, \tag{7.4.1}$$

where

$$\Phi = \frac{1}{2\pi} \log|z - c| - \frac{1}{4\pi} \mathscr{R}\{\Omega(z, c) - \Omega(z, \bar{c})\}, \tag{7.4.2}$$

where $c = a + ib$ and

$$\Omega(z, c) = \log(z - c) - \log\left[\frac{z + (z^2 - 1)^{1/2} - c - (c^2 - 1)^{1/2}}{z + (z^2 - 1)^{1/2} - c + (c^2 - 1)^{1/2}}\right] - \cosh^{-1}z, \tag{7.4.3}$$

$$\Gamma = \psi_{11}n_1 + \psi_{12}n_2, \tag{7.4.4}$$

where

$$\psi_{11} = \frac{1}{2\pi} \mathscr{R}\left\{\frac{1}{z - c}\right\} - \frac{1}{4\pi} \mathscr{R}\{(z^2 - 1)^{-1/2}[\chi(z, c) - \chi(z, \bar{c})]\}, \tag{7.4.5}$$

$$\psi_{12} = \frac{1}{2\pi} \mathscr{R}\left\{\frac{i}{z - c}\right\} - \frac{1}{4\pi} \mathscr{R}\{i(z^2 - 1)^{-1/2}[\chi(z, c) - \chi(z, \bar{c})]\}, \tag{7.4.6}$$

with

$$\chi(z, c) = \frac{(z^2 - 1)^{1/2} - (c^2 - 1)^{1/2}}{z - c} - 1. \tag{7.4.7}$$

It is of interest to note that, in the case of Laplace's equation, the final terms in (7.4.3) and (7.4.7) do not enter into the expressions for Φ and Γ.

In (7.4.1) let $\phi = 1$ (note that this does not violate the condition under which (7.4.1) was obtained; namely that $P = \partial\phi/\partial n = 0$ over the cut) to obtain

$$\lambda = \int_{C_2} \Gamma(\mathbf{x}, \mathbf{x}_0) \, ds(\mathbf{x}), \tag{7.4.8}$$

and substituting this into (7.4.1) yields the integral equation

$$\int_{C_2} P(\mathbf{x})\phi(\mathbf{x}, \mathbf{x}_0) \, ds(\mathbf{x}) = \int_{C_2} \Gamma(\mathbf{x}, \mathbf{x}_0)[\phi(\mathbf{x}) - \phi(\mathbf{x}_0)] \, ds(\mathbf{x}). \tag{7.4.9}$$

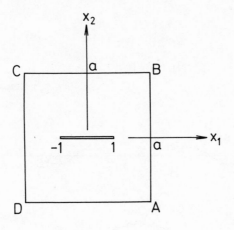

Fig. 7.4.1

This equation may be solved numerically by using a similar procedure to the one used in Section 7.2 for equation (7.2.9). The expressions for Φ and Γ are more complicated in this case since they involve extra terms. However, these extra terms are analytic along C_2 and hence may be readily integrated using simple quadrature formulae. The only minor difficulty concerns the square root. Care must be taken to ensure that the branch of the square root is such that $z^{-1}(z^2-1)^{1/2} \to 1$ as $|z| \to \infty$.

As an example consider the problem discussed by Clements and King [13]. This concerns finding a solution to Laplace's equation for the domain R with boundary $C_1 + C_2$ shown in Fig. 7.4.1. The boundary conditions are

$$P = \frac{\partial \phi}{\partial n} = 0 \quad \text{on} \quad C_1,$$

$$\phi = \mathscr{R}[i(a + ix_2 - 1)^{1/2}] \quad \text{on} \quad AB,$$

$$\phi = \mathscr{R}[i(x_1 + ia) - 1)^{1/2}] \quad \text{on} \quad BC,$$

$$\phi = \mathscr{R}[i(-a + ix_2 - 1)^{1/2}] \quad \text{on} \quad CD,$$

$$\phi = \mathscr{R}[i(x_1 - ia - 1)^{1/2}] \quad \text{on} \quad DA.$$

(7.4.10)

For this problem equation (7.4.9) is applicable and may be used to find P at various points on C_2. Equation (7.4.1) then gives ϕ (and by differentiation P) at all interior points of the domain R. Here it will be sufficient to use (7.4.9) to obtain numerical values for P on C_2 and then to compare these results with those obtained from the analytical solution

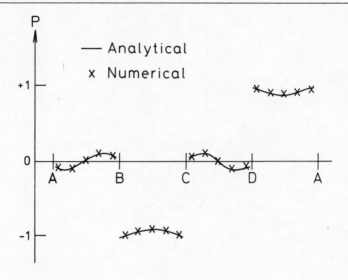

Fig. 7.4.2

to this problem. This solution takes the form

$$\phi = \mathcal{R}[i(z^2-1)^{1/2}], \tag{7.4.11}$$

$$P = \frac{\partial \phi}{\partial x_1} n_1 + \frac{\partial \phi}{\partial x_2} n_2$$

$$= \mathcal{R}\left[\left(\frac{iz}{(z^2-1)^{1/2}}\right)n_1 - \left(\frac{z}{(z^2-1)^{1/2}}\right)n_2\right]. \tag{7.4.12}$$

For illustrative purposes let $a = 2$. In Fig. 7.4.2 the continuous lines represent the analytic solution for P and the crosses the numerical values obtained by using five segments on each side (that is, $N = 20$). The boundary is stretched out to form the horizontal axis. As can be seen from the graph, the accuracy obtained, even when using only a small number of segments, is extremely good. If the number of segments is increased, for example, to forty, then the analytical and numerical solutions are indistinguishable on the graph.

To illustrate the accuracy of the numerical solution the values are tabulated for twenty and forty segments. The values on the sides AB and BC only are given as the symmetry of the solution may be used to determine P on the other sides of the region (see Fig. 7.4.1). The percentage error shown in Tables 7.4.1 and 7.4.2 is given by

$$\text{Percentage error} = 100 \times \left|\frac{\text{analytical solution} - \text{numerical solution}}{\text{analytical solution}}\right|.$$

Table 7.4.1 Comparison of analytical and numerical solutions for twenty segments

	Position (x, y)	Analytical solution	Numerical solution	Percentage error
AB	2.0, −1.6	−0.0771	−0.0689	10.6
	2.0, −0.8	−0.0940	−0.0990	5.3
	2.0, 0.0	0.0	0.000	0
	2.0, 0.8	0.0940	0.0990	5.3
	2.0, 1.6	0.0771	0.0689	10.6
BC	1.6, 2.0	−0.9762	−0.9752	0.1
	0.8, 2.0	−0.9244	−0.9228	0.17
	0.0, 2.0	−0.8944	−0.8922	0.25
	−0.8, 2.0	−0.9244	−0.9228	0.17
	−1.6, 2.0	−0.9762	−0.9752	0.1

Tables 7.4.1 and 7.4.2 show that the errors on the side AB are much larger than those on BC with the largest error occurring at the ends of AB. By comparing the two tables it can be seen that in general the accuracy of the numerical approximation is significantly improved by taking more segments. The only segments for which this is not true are the end segments of AB. The percentage errors on AB are the larger as

Table 7.4.2 Comparison of analytical and numerical solutions for forty segments

	Position (x, y)	Analytical solution	Numerical solution	Percentage error
AB	2.0, −1.8	−0.0694	−0.0616	11.2
	2.0, −1.4	−0.0847	−0.0868	2.48
	2.0, −1.0	−0.0950	−0.0957	0.74
	2.0, −0.6	−0.0855	−0.0867	1.4
	2.0, −0.2	−0.0371	−0.0377	1.6
	2.0, 0.2	0.0371	0.0377	1.6
	2.0, 0.6	0.0855	0.0867	1.4
	2.0, 1.0	0.0950	0.0957	0.74
	2.0, 1.4	0.0847	0.0868	2.48
	2.0, 1.8	0.0694	0.0616	11.2
BC	1.8, 2.0	−0.9861	−0.9842	0.19
	1.4, 2.0	−0.9645	−0.9641	0.04
	1.0, 2.0	−0.9378	−0.9374	0.04
	0.6, 2.0	−0.9123	−0.9117	0.07
	0.2, 2.0	−0.8966	−0.8958	0.1
	−0.2, 2.0	−0.8966	−0.8958	0.1
	−0.6, 2.0	−0.9123	−0.9117	0.07
	−1.0, 2.0	−0.9378	−0.9374	0.04
	−1.4, 2.0	−0.9645	−0.9641	0.04
	−1.8, 2.0	−0.9861	−0.9842	0.19

on this side the numbers being calculated are much smaller than those on *BC*. However, the actual errors on all sides are of approximately the same order.

It is of interest to examine the effect of a cut by comparing the results for a plane with a cut and one without a cut. To do this, we consider the square *ABCD* shown in Fig. 7.2.1 with the boundary conditions

$$\phi = 2 + x_2 \quad \text{on} \quad AB,$$
$$\phi = 2 + x_1 \quad \text{on} \quad BC,$$
$$\phi = -2 + x_2 \quad \text{on} \quad CD,$$
$$\phi = -2 + x_1 \quad \text{on} \quad DA,$$

(7.4.13)

and with $P = \partial\phi/\partial n$ on the cut if the region has one. Firstly, if the region has no cut then the boundary conditions (7.4.13) allow the analytical solution $\phi = x_1 + x_2$ everywhere in the region. Hence

$$P = \frac{\partial\phi}{\partial n} = \begin{cases} 1 & \text{on} \quad AB \\ 1 & \text{on} \quad BC \\ -1 & \text{on} \quad CD \\ -1 & \text{on} \quad DA. \end{cases}$$

(7.4.14)

These values of P are shown as the continuous lines in Fig. 7.4.3. In the case when the region contains a cut as shown in Fig. 7.4.1 then, with the boundary conditions (7.4.13) on *ABCD*, the problem does not admit a simple solution. The numerical solution is given by the broken lines. As

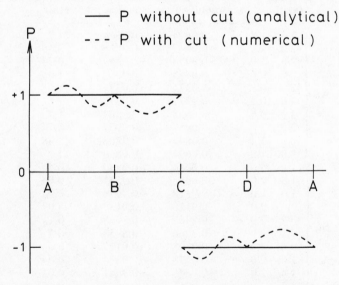

Fig. 7.4.3

can be seen from the graph, the presence of the cut does have a significant effect on P for this particular problem.

7.5 Some problems in anisotropic thermostatics

The equation governing the temperature field in generalized plane thermostatics is (see Section 2.3)

$$\lambda_{ij}\frac{\partial^2 T}{\partial x_i\,\partial x_j}=0. \tag{7.5.1}$$

Here the equations derived in Chapter 6 are used to obtain numerical solutions to this equation which are valid in a region R in E^2 with boundary C. On C either ϕ is specified or the heat flux

$$P=\lambda_{ij}\frac{\partial T}{\partial x_j}\,n_i \tag{7.5.2}$$

is given with n_i the unit outward normal to R. If P is specified at all points of C then it must be such that

$$\int_C P\,\mathrm{d}s=0. \tag{7.5.3}$$

The boundary integral equation for the solution of problems of this type is (6.3.4) with Φ_{ij} and Γ_{ij} given by (6.3.5)–(6.3.7) where the constants occurring in these equations are given by (2.3.7). Using (2.3.7) in (6.3.4)–(6.3.7) and dropping the subscripts which are unnecessary in this case it follows that (6.3.4) may be written (with the arbitrary constant $F=\lambda_{11}^{-1}$ and $\phi=T$) as

$$\lambda T(\mathbf{x}_0)+\lambda_{11}^{-1}\int_C [P(\mathbf{x})\Phi(\mathbf{x},\mathbf{x}_0)-\Gamma(\mathbf{x},\mathbf{x}_0)T(\mathbf{x})]\,\mathrm{d}s(\mathbf{x})=0, \tag{7.5.4}$$

where

$$\Phi=\frac{-1}{2\pi i(\tau-\bar\tau)}\left(\frac{\lambda_{11}}{\lambda_{22}}\right)\{\log(z_1-c)+\log(\bar z_1-\bar c)\}, \tag{7.5.5}$$

$$\Gamma=\left\{\frac{\mathcal{M}}{z_1-c}+\frac{\bar{\mathcal{M}}}{\bar z_1-\bar c}\right\}, \tag{7.5.6}$$

with $z_1=x_1+\tau x_2$, $c=a+\tau b$ and τ is the root with positive real part of the equation

$$\lambda_{11}+2\lambda_{12}\tau+\lambda_{22}\tau^2=0. \tag{7.5.7}$$

Also, in (7.5.6)

$$\mathcal{M} = \frac{-1}{2\pi i(\tau - \bar{\tau})} \left(\frac{\lambda_{11}}{\lambda_{22}}\right)\{\lambda_{11}n_1 + \lambda_{12}n_2 + (\lambda_{12}n_1 + \lambda_{22}n_2)\tau\}. \qquad (7.5.8)$$

Consider an anisotropic material for which $\lambda_{22}/\lambda_{11} = 1$ and $\lambda_{12}/\lambda_{11} = 0.5$. As a test problem let the domain R consist of the square $ABCD$ as shown in Fig. 7.2.1. The boundary conditions are taken to be the same as those considered in Section 7.2. Specifically,

$$
\begin{aligned}
T(0.6, x_2) &= 0.6 + x_2 & \text{on} \quad AB, \\
T(x_1, 0.6) &= x_1 + 0.6 & \text{on} \quad BC, \\
T(-0.6, x_2) &= -0.6 + x_2 & \text{on} \quad CD, \\
T(x_1, -0.6) &= x_1 - 0.6 & \text{on} \quad DA.
\end{aligned}
\qquad (7.5.9)
$$

The analytical solution to this problem is $T = x_1 + x_2$ so that

$$P = 1.5(n_1 + n_2). \qquad (7.5.10)$$

Equation (7.5.4) may be used to determine numerical values of P on the boundary C. The procedure is very similar to that described in Section 7.2. The only difference of any significance is in the formulae for a_{mj}. Firstly consider the case when $m = j$. By integrating $\log|z_1 - c|$ along the straight lines joining $\mathbf{q}_{m-1}$ to $\bar{\mathbf{q}}_m$ and $\bar{\mathbf{q}}_m$ to $\mathbf{q}_m$ it may readily be verified that

$$
\int_{\mathbf{q}_{m-1}}^{\mathbf{q}_m} \log|z_1 - c|\, ds(\mathbf{x})
$$

$$
= |\bar{\mathbf{q}}_m - \mathbf{q}_{m-1}| \{-1 + \log|\bar{x}_m - x_{m-1} + \tau(\bar{y}_m - y_{m-1})|\}
$$
$$
+ |\mathbf{q}_m - \bar{\mathbf{q}}_m| \{-1 + \log|x_m - \bar{x}_m + \tau(y_m - \bar{y}_m)|\}, \quad (7.5.11)
$$

where $\mathbf{q}_m = (x_m, y_m)$, $\bar{\mathbf{q}}_m = (\bar{x}_m, \bar{y}_m)$ and $c = \bar{x}_m + \tau\bar{y}_m$. Hence

$$
a_{mm} = \int_{\mathbf{q}_{m-1}}^{\mathbf{q}_m} \Phi(x, \bar{\mathbf{q}}_m)\, ds(x)
$$

$$
= \frac{-1}{\pi i(\tau - \bar{\tau})} \left(\frac{\lambda_{11}}{\lambda_{22}}\right)[|\bar{\mathbf{q}}_m - \mathbf{q}_{m-1}| \{-1 + \log|\bar{x}_m - x_{m-1} + \tau(\bar{y}_m - y_{m-1})|\}
$$

$$
+ |\mathbf{q}_m - \bar{\mathbf{q}}_m| \{-1 + \log|x_m - \bar{x}_m + \tau(y_m - \bar{y}_m)|\}. \qquad (7.5.12)
$$

If $j \neq m$ so that $\bar{\mathbf{q}}_j$ is not on the segment from $\mathbf{q}_{m-1}$ to $\mathbf{q}_m$, then Simpson's rule yields

$$
a_{mj} = \frac{h_m}{3} \left(\frac{-1}{\pi i(\tau - \bar{\tau})}\right)\left(\frac{\lambda_{11}}{\lambda_{22}}\right)\{\log|x_{m-1} - \bar{x}_j + \tau(y_{m-1} - \bar{y}_j)|
$$

$$
+ 4\log|\bar{x}_m - \bar{x}_j + \tau(\bar{y}_m - \bar{y}_j)| + \log|x_m - \bar{x}_j + \tau(y_m - \bar{y}_j)|\}. \quad (7.5.13)
$$

Table 7.5.1

Point	$P(\mathbf{x})/\lambda_{11}$ analytical	$P(\mathbf{x})/\lambda_{11}$ numerical	% Error
q_1	−1.5	−1.516	1.6
q_2	−1.5	−1.494	0.6
q_3	−1.5	−1.500	0
q_4	−1.5	−1.500	0
q_5	−1.5	−1.494	0.6
q_6	−1.5	−1.516	1.6
q_7	1.5	1.516	1.6
q_8	1.5	1.494	0.6
q_9	1.5	1.500	0
q_{10}	1.5	1.500	0
q_{11}	1.5	1.494	0.6
q_{12}	1.5	1.516	1.6

As in Section 7.2, the numerical values of P on each segment of the boundary may be calculated from (7.2.12) where, in this case, the a_{mj} are given by (7.5.12) and (7.5.13) and the R_j are obtained from (7.2.14) with Γ given by (7.5.6) and (7.5.7).

So that the results may be compared with those of Section 7.2 it is useful to divide each side of the square into three equal segments as shown in Fig. 7.2.1. Values of P_m for $m = 1, 2, \ldots, 12$ obtained by using (7.5.4) are given in Table 7.5.1 together with those obtained from the analytical solution (7.5.10). The numerical and analytical results differ by less than 2%. A comparison of Tables 7.2.1 and 7.5.1 shows that the maximum error is slightly less in the case of Laplace's equation. However, taken overall, there is negligible difference in the accuracy of the results for Laplace's equation and equation (7.5.1) with $\lambda_{22}/\lambda_{11} = 1$ and $\lambda_{12}/\lambda_{11} = 0.5$.

As a second example consider the case when the domain R consists of the unit circle (Fig. 7.2.3) and the boundary conditions are

$$T = \cos \theta + \sin \theta,$$

so that the analytical solution to the problem is $T = x_1 + x_2$. Hence if $\lambda_{22}/\lambda_{11} = 1$ and $\lambda_{12}/\lambda_{11} = 0.5$ then P is again given by (7.5.10). The boundary is divided into N equal segments as described in Section 7.2. With $N = 20$ equation (7.5.4) yields values of P which differ from the analytical values by approximately 10%. For $N = 40$ the numerical and analytical values for some particular boundary points are given in Table 7.5.2. Note that although the agreement between the analytical and numerical results is excellent it is not as good as the agreement obtained in Section 7.2 for Laplace's equation (see Table 7.2.2).

Table 7.5.2

Boundary point $(\cos\theta, \sin\theta)$	$P(\mathbf{x})/\lambda_{11}$ analytical	$P(\mathbf{x})/\lambda_{11}$ numerical	% Error
$\theta = 270°$	-1.4978	-1.5	0.22
$\theta = 279°$	-1.2427	-1.2468	0.41
$\theta = 288°$	-0.9567	-0.9630	0.63
$\theta = 297°$	-0.6478	-0.6555	0.77
$\theta = 306°$	-0.3261	-0.3318	0.57
$\theta = 315°$	0	0	0
$\theta = 324°$	0.3261	0.3318	0.57
$\theta = 333°$	0.6478	0.6555	0.77
$\theta = 342°$	0.9567	0.9630	0.63
$\theta = 351°$	1.2427	1.2468	0.41
$\theta = 360°$	1.4978	1.5	0.22

Consider next a problem examined by Schulock [33]. This problem involves using (7.5.4) to solve (7.5.1) for the case when the region R consists of a square with the following boundary conditions:

$$
\begin{aligned}
T &= 2T_0 & \text{on } x_2 = 1, \\
P &= \lambda_{12}T_0 & \text{on } x_1 = 1, \\
T &= 0 & \text{on } x_2 = -1, \\
P &= -\lambda_{12}T_0 & \text{on } x_1 = 1,
\end{aligned}
\tag{7.5.14}
$$

where T_0 is a constant. Also

$$
\lambda_{11}/\lambda = \lambda_{22}/\lambda = \tfrac{1}{2}(\alpha + 1), \qquad \lambda_{12}/\lambda = \tfrac{1}{2}(\alpha - 1),
$$

where α is a parameter and λ is a reference coefficient of heat conduction. This problem admits the simple analytical solution

$$
T = T_0(x + 1), \tag{7.5.15}
$$

so that

$$
P = T_0(\lambda_{11}n_1 + \lambda_{12}n_2). \tag{7.5.16}
$$

The unknown temperatures or fluxes were calculated by Schulock using one segment per side and a 4-point Legendre Gauss rule for numerical quadrature. The average absolute error ε_Ψ for the calculation is shown in Table 7.5.3 for various values of α. The ε_Ψ is defined by

$$
\varepsilon_\Psi = \frac{1}{M_\Psi} \frac{\sqrt{(\Psi_{\text{NUMERICAL}} - \Psi_{\text{EXACT}})^2}}{\Psi_{\text{EXACT}}} \tag{7.5.17}
$$

Table 7.5.3 Average absolute error $\varepsilon_\Psi \times 10^4$

α	2	3	4	5	6	7	8	9	10	11	12	13	14	15
Temp.	2	2	7	8	3	7	24	47	76	109	147	188	234	282
Flux	12	16	18	24	28	33	52	73	93	114	113	174	176	196

where Ψ is either variable being solved for in the mixed problem (that is, the temperature or the flux) and M_Ψ is the number of nodes at which the numerical value of Ψ is obtained. It is clear from Table 7.5.3 that ε_Ψ increases with α. This is not altogether surprising since when α becomes large, equation (7.5.1) becomes weakly elliptic. As has been indicated in previous chapters, this situation often arises in strongly anisotropic or fibre-reinforced materials. Thus the present method could be used to solve problems for this important class of materials since, even though the error increases with α, it is still small in magnitude.

7.6 Some problems in anisotropic elasticity

The problem to be considered in this section is to find displacements u_1, u_2 and u_3 which satisfy the equations of elastic equilibrium

$$c_{ijkl}\frac{\partial^2 u_k}{\partial x_j\,\partial x_1} = 0 \quad \text{for} \quad i = 1, 2, 3 \tag{7.6.1}$$

in a region R in E^2 with boundary C. On C either the displacements u_1, u_2 and u_3 are specified or the tractions

$$P_i = c_{ijkl}\frac{\partial u_k}{\partial x_1}\,n_j \quad \text{for} \quad i = 1, 2 \tag{7.6.2}$$

are given with n_i the unit outward normal to R. If the tractions are specified at all points of C they cannot be specified completely arbitrarily but are subject to the condition (1.1.6).

The boundary integral equation for the solution of this problem is (6.3.4) with Φ_{ij} and Γ_{ij} given by (6.3.5)–(6.3.7) where the constants occurring in these equations are given by (2.4.10)–(2.4.14). Equation (6.3.4) may be used numerically to solve a particular boundary-value problem by employing straightforward generalizations of the techniques outlined in Sections 7.2 by Laplace's equation.

Some particular boundary-value problems have been solved by Rizzo and Shippy [31] who considered orthotropic elastic materials so that the only non-zero elastic constants are given by (2.7.18). Hence, written out in full, equation (7.6.1) yields

$$c_{1111} \frac{\partial^2 u_1}{\partial x_1^2} + (c_{1122} + c_{1212}) \frac{\partial^2 u_2}{\partial x_1 \partial x_2} + c_{1212} \frac{\partial^2 u_1}{\partial x_2^2} = 0,$$

$$c_{2222} \frac{\partial^2 u_2}{\partial x_2^2} + (c_{1122} + c_{1212}) \frac{\partial^2 u_1}{\partial x_1 \partial x_2} + c_{1212} \frac{\partial^2 u_2}{\partial x_1^2} = 0,$$

(7.6.3)

$$c_{3131} \frac{\partial^2 u_3}{\partial x_1^2} + c_{3232} \frac{\partial^2 u_3}{\partial x_2^2} = 0.$$

(7.6.4)

In this case the plane and antiplane parts of the problem uncouple. The plane part involves the displacements u_1 and u_2 and is governed by (7.6.3) while the antiplane part involves the single displacement u_3 whose behaviour is governed by equation (7.6.4). Rizzo and Shippy only considered plane problems and hence the displacement u_3 does not enter into their analysis. Also they used material constants s_{11}, s_{12}, s_{22} and s_{66} which are related to the constants occurring in (7.6.3) by the equations

$$c_{1111} = s_{22}/D, \qquad c_{1122} = -s_{12}/D, \qquad c_{2222} = s_{11}/D,$$

$$c_{1221} = 1/s_{66}, \qquad D = s_{11}s_{22} - s_{12}^2.$$

(7.6.5)

The first problem considered by Rizzo and Shippy concerns an elliptic inclusion in an infinite material. The problem involves forcing an elliptic plug of semi-axes $a(1+\varepsilon)$, $b(1+\varepsilon)$ into a slightly smaller elliptical hole with semi-axes a, b. Using an equation of the type (6.3.4) together with straightforward extensions of the numerical procedure outlined in Section 7.2, Rizzo and Shippy calculated values of the stress at various points in the material. For the purpose of the calculations it is assumed that

$$s_{11} = s_{22}, \qquad s_{12} = -\tfrac{1}{2}s_{11}, \qquad a/b = 2 \qquad 3s_{11}/s_{66} = \tfrac{1}{8}.$$

Also the major and minor axes of the elliptical hole and inclusion are required to be lined up with the Cartesian axes $0x_1$, $0x_2$ with $a = 20\,000$ in, $b = 10\,000$ in, $\varepsilon = 2 \times 10^{-4}$ and $s_{11} = 10^{-7}$ in.^{2}/lb. The stresses

Table 7.6.1

Coordinates (in × 10^4)		Stresses (psi × 10^3)	
x_1	x_2	σ_{11}	σ_{22}
0	0	−0.634	−0.234
0.8	0	−0.634	−0.234
1.6	0	−0.657	−0.232
0	0.2	−0.633	−0.234
0	0.6	−0.634	−0.240
0	1.0	−0.636	−0.234

σ_{11} and σ_{22} are uniform throughout the inclusion at $-610\,\text{psi}$ and $-228\,\text{psi}$, respectively, and σ_{12} is zero everywhere in the inclusion. These analytical values may be compared with those obtained numerically and listed in Table 7.6.1. These values were obtained by using an equation of the type (6.3.4) with the boundary divided into 48 equal intervals.

It is apparent that the numerical and analytical results are in excellent agreement. As well as carrying out further calculations on this particular problem Rizzo and Shippy also considered the problem of a circular hole in an infinite plate and a mixed problem for an orthotropic ring. In each case the numerical results were shown to be in excellent agreement with those obtained from known analytical solutions.

Bibliography

1 Abramowitz, M. and Stegun, L. A., *Handbook of mathematical functions*, (Dover), New York, 1965.

2 Atkinson, C., The interaction between a dislocation and a crack, *Int. J. Fracture Mech.*, **2** (1966), 567–575.

3 Atkinson, C. and Head, A. K., The influence of elastic anisotropy on the propagation of fracture, *Int. J. Fracture Mech.*, **2** (1966), 377–390.

4 Brilla, J., Contact problems of an elastic anisotropic half-plane, *Revue de Mec. Appl.*, **7** (1962), 617–644.

5 Carslaw, H. S. and Jaeger, J. C., *Conduction of heat in solids*, (Clarendon Press), Oxford, 1959.

6 Clements, D. L., A crack between dissimilar anisotropic media, *Int. J. Engng. Sci.*, **9** (1971), 257–265.

7 Clements, D. L., The motion of a heavy cylinder over the surface of an anisotropic elastic solid, *J. Inst. Maths. Applics.*, **7** (1971), 198–206.

8 Clements, D. L., A property of contact problems for anisotropic and fiber-reinforced half-spaces, *Utilitas Mathematica*, **2** (1972), 33–46.

9 Clements, D. L., Two contact problems in anisotropic elasticity, *J. Aust. Math. Soc.*, **15** (1973), 35–41.

10 Clements, D. L., The effect of an axial force on the response of an anisotropic elastic half-space to a rolling cylinder, *J. Appl. Mech.*, **40** (1973), 251–256.

11 Clements, D. L., A crack in an anisotropic elastic slab, *Quart. Appl. Math.*, **34** (1977), 437–443.

12 Clements, D. L. and Rizzo, F. J., A method for the numerical solution of boundary value problems governed by second-order elliptic systems, *J. Inst. Maths. Applics.*, **22** (1978), 197–202.

13 Clements, D. L. and King, G. W., A method for the numerical solution of problems governed by elliptic systems in the cut plane, *J. Inst. Maths. Applics.*, **24** (1979), 81–93.

14 Clements, D. L., A crack in an anisotropic layered material, *Engng. Transactions*, **27** (1979), 171–180.

15 Cruse, T. A. and Rizzo, F. J. (Eds.), Boundary integral equation method: computational applications in applied mechanics, ASME Proceedings, AMD Vol. II, 1975.

16 England, A. H., A crack between dissimilar media, *J. Appl. Mech.*, **32** (1965), 400–402.

17 England, A. H., *Complex variable methods in elasticity*, (Wiley), London, 1971.

18 Eshelby, J. D., Read, W. T. and Shockley, W., Anisotropic elasticity with applications to dislocation theory, *Acta Met.*, **1** (1953), 251–259.

19 Everstine, G. C. and Pipkin, A. C., Stress channelling in transversely isotropic elastic composites, *Z. Angew. Math. Phys.*, **22** (1971), 825–834.

20 Green, A. E. and Zerna, W., *Theoretical elasticity*, (Oxford University Press), London, 1954.

21 Griffith, A. A., The phenomenon of rupture and flow in solids, *Phil. Trans. Roy. Soc.*, **A221** (1921), 163–198.

22 John, F., *Plane waves and spherical means*, (Interscience), New York, 1955.

23 Jaswon, M. A., Integral equation methods in potential theory, *Proc. Roy. Soc.*, **A275** (1963), 23–32.

24 Jaswon, M. A. and Ponter, A. R. S., An integer equation solution of the torsion problem, *Proc. Roy. Soc.*, **A273** (1963), 237–246.

25 Jaswon, M. A. and Symm, G. T., *Integral equation methods in potential theory and elastostatics*, (Academic Press), London, 1977.

26 Lekhnitskii, S. G., *Anisotropic plates*, (Gordon and Breach), New York, 1968.

27 Lekhnitskii, S. G., *Theory of elasticity of an anisotropic elastic body*, (Holden-Day), San Francisco, 1963.

28 Muskhelishvili, N. I., *Singular integral equations*, (Noordhoff), Holland, 1953.

29 Muskhelishvili, N. I., *Some basic problems in the mathematical theory of elasticity*, (Noordhoff), Holland, 1953.

30 Pipkin, A. C. and Rogers, T. G., Plane deformations of incompressible fiber-reinforced materials, *J. Appl. Mech.*, **38** (1971), 634–640.

31 Rizzo, F. J. and Shippy, D. J., A method for stress determination in plane anisotropic elastic bodies, *J. Composite Mats.*, **4** (1970), 36–61.

32 Salganik, R. A., The brittle fracture of cemented bodies, *J. Appl. Math. Mech.*, **27** (1963), 1468–1478.

33 Schulock, M. L., The boundary integral equation method for the numerical solution of boundary value problems governed by second order elliptic systems, M.Sc. Thesis, University of Kentucky (1978).

34 Sneddon, I. N., *Fourier transforms*, (McGraw-Hill), New York, 1951.

35 Snyder, M. D. and Cruse, T. A., Boundary-integral equation analysis of cracked anisotropic plates, *Int. J. Fract.*, **11** (1975), 315–328.

36 Spencer, A. J. M., *Deformations of fibre-reinforced materials*, (Clarendon Press), Oxford, 1972.

37 Stakgold, I., *Boundary value problems of mathematical physics*, Vols. I & II, (Macmillan), New York, 1971.

38 Stroh, A. N., Dislocations and cracks in anisotropic elasticity, *Phil. Mag.*, **3** (1958), 625–646.

39 Symm, G. T., Integral equation methods in potential theory II, *Proc. Roy. Soc.*, **A275** (1963), 33–46.

40 Tauchert, T. R., Stresses in an anisotropic elastic slab due to distributed surface loads and displacements, in *Developments in mechanics*, Vol. 8. Proceedings of the 14th Midwestern Mechanics Conference (1975), 89–101.

41 Willis, J. R., Fracture mechanics of interfacial cracks, *J. Mech. Phys. Solids*, **19** (1971), 353–368.

42 Zienkiewicz, O. C., Kelly, D. W. and Bettess, P., The coupling of the finite element method and boundary solution procedures, *Int. J. Num. Methods in Engng.*, **11** (1977), 355–375.

Index